Cambridge Elements ≡

Elements in the Philosophy of Biology
edited by
Grant Ramsey
KU Leuven
Michael Ruse
Florida State University

PHILOSOPHY OF DEVELOPMENTAL BIOLOGY

Marcel Weber
University of Geneva

CAMBRIDGE
UNIVERSITY PRESS

CAMBRIDGE
UNIVERSITY PRESS

University Printing House, Cambridge CB2 8BS, United Kingdom

One Liberty Plaza, 20th Floor, New York, NY 10006, USA

477 Williamstown Road, Port Melbourne, VIC 3207, Australia

314–321, 3rd Floor, Plot 3, Splendor Forum, Jasola District Centre, New Delhi – 110025, India

103 Penang Road, #05–06/07, Visioncrest Commercial, Singapore 238467

Cambridge University Press is part of the University of Cambridge.

It furthers the University's mission by disseminating knowledge in the pursuit of education, learning, and research at the highest international levels of excellence.

www.cambridge.org
Information on this title: www.cambridge.org/9781009184151
DOI: 10.1017/9781108954181

First published 2022

A catalogue record for this publication is available from the British Library.

ISBN 978-1-009-18415-1 Hardback
ISBN 978-1-108-94935-4 Paperback
ISSN 2515-1126 (online)
ISSN 2515-1118 (print)

Philosophy of Developmental Biology

Elements in the Philosophy of Biology

DOI: 10.1017/9781108954181
First published online: March 2022

Marcel Weber
University of Geneva

Author for correspondence: Marcel Weber, marcel.weber@unige.ch

Abstract: The history of developmental biology is interwoven with debates as to whether mechanistic explanations of development are possible or whether alternative explanatory principles or even vital forces need to be assumed. In particular, the demonstrated ability of embryonic cells to tune their developmental fate precisely to their relative position and the overall size of the embryo was once thought to be inexplicable in mechanistic terms. Taking a causal perspective, this Element examines to what extent and how developmental biology, having turned molecular about four decades ago, has been able to meet the vitalist challenge. It focuses not only on the nature of explanations but also on the usefulness of causal knowledge – including the knowledge of classical experimental embryology – for further scientific discovery. It also shows how this causal perspective allows us to understand the nature and significance of some key concepts, including organizer, signal and morphogen. This title is also available as Open Access on Cambridge Core.

Keywords: developmental biology, philosophy of biology, causality, mechanism, vitalism

ISBNs: 9781009184151 (HB), 9781108949354 (PB), 9781108954181 (OC)
ISSNs: 2515-1126 (online), 2515-1118 (print)

Contents

1 Introduction: From Mechanism to Vitalism and Back

This section provides a brief historical and philosophical framing for the central issues to be discussed in this Element, which have to do with experiment, causality, and explanation in developmental biology before and after its molecular turn. I begin by telling the story of a puzzling discovery in experimental embryology from the late nineteenth century, a phenomenon known as embryonic regulation or regulative epigenesis. Its discoverer, Hans Driesch, later came to believe that it could not be explained mechanistically and was thus led to postulate an immaterial vital force. His arguments for vitalism didn't win many supporters, yet his influence on biological thought was considerable. In any case, the phenomena discovered by Driesch and their possible mechanistic and molecular basis have engaged developmental biologists ever since. Following their fate into contemporary developmental biology provides insight into the workings of experimental science.

1.1 A Puzzling Discovery

Developmental biology today studies a vast range of biological processes that occur in animal and plant embryos as well as in adult organisms, including gamete formation and fertilization, embryonic pattern formation, cell differentiation, organogenesis, limb formation, regeneration, senescence and aging, as well as evolution (e.g., Gilbert and Barresi, 2016).[1] The historical origins of this science, which only emerged as an independent professional discipline in the 1930s and 1940s, are diverse and include experimental as well as anatomical and comparative research traditions (Hopwood, 2009). The experimental tradition began to flourish in the nineteenth century with a research program that was best known at the time as *Entwicklungsmechanik,* which is German for "developmental mechanics," but which was also referred to as "physiological embryology" or "causal embryology." Its advent is usually described as a turn from a natural history-based approach to an experimental science that seeks to identify the causes of embryonic development (Maienschein, 1991). Indeed, the hallmark of *Entwicklungsmechanik* was a thoroughly experimental approach. For example, one of its chief proponents, Wilhelm Roux, punctured single cells in frog embryos with a needle and obtained half frog embryos. He used this result to support his mosaic theory of development according to which embryonic cells divide unevenly such that their daughter cells will rigidly develop into different parts of the body.

---

[1] Evolutionary developmental biology or "evo-devo" is treated in a different Element by Alan Love.

It was later shown that Roux's results were actually due to the dead cells that remained attached to the manipulated embryos. Normally, rather than maintaining a rigid determination of their fate, embryonic frog cells in fact remain responsive to outside signals that adapt their fate to their location within the embryo, at least during a certain time window. Unlike what Roux sought to prove, frog development is an example of what was referred to as "epigenesis," which could be defined as the generation of new structure due to interactions between different parts of the embryo. By contrast, Roux's mosaic model rather looked like the unfolding of structures that preexisted in some of parts of the embryo, as so-called "preformationist" theories of development held.

Roux's experiments stimulated a lot of research that eventually firmly established the reality of epigenesis, also in frogs. But Roux's work was at least as important in promoting a specific approach: by experimentally manipulating embryos of a model organism used as a stand-in for other organisms (Ankeny and Leonelli, 2020), he sought to learn something about the dispositions of embryonic cells.

While the experimental approach to development was initially associated with mechanistic doctrines, this was soon met with resistance. A keen experimenter with a liking of German philosophy (especially Kant), Hans Driesch was well-known for his experimental work on sea urchins that seemed to be at odds with Roux's earlier frog findings. In fact, Driesch's sea urchin embryos looked like the exact opposite of Roux's frogs: when he separated the embryo at the two- or four-cell stage of development with a hairpin, each one of the cells formed a whole and perfectly happy (although somewhat smaller) sea urchin larva. Driesch argued that this result supported a "regulative epigenesis" rather than Roux's mosaic theory. (It was later found that frog embryos can do the same if the cells are properly detached rather than just punctured; see Maienschein, 1991: 50). But Driesch soon went further. He showed that some parts of the sea urchin embryo retain the potential to form a whole organism up to the 800-cell stage. In addition to his results with sea urchin embryos, Driesch was also experimenting with adult marine organisms that showed remarkable regenerative powers. For example, he showed that sea squirts of the genus *Clavellina* were able to regenerate large parts of the body after surgical removal, and that some parts (such as the branchial syphon) were even able to form a whole new sea squirt.

Driesch suggested that all these systems had something in common, namely, they formed what he called "harmonious equipotential systems." Roughly, these are systems in which each part has the same potential as all the other parts and also the same potential as the whole. For example, all the cells in a four-cell sea urchin embryo have the same potential as the other cells and the same potential as the whole, namely the potential to form a whole organism. The cell's actual fate (i.e., the structures it ends up actually forming by a series of cell divisions and movements) depends on the cell's location and on the size of the embryo. To explain this, Driesch postulated a factor "*E*" that is responsive to the cell's location and the embryo's size and which tells the cells what they should become. Formally, the fate *S* of any part of a harmonious-equipotential system can be written as a function: $S = f(a, g, E)$, where *a* is a position vector, *g* a scalar expressing the size of the system, and *E* the factor that Driesch called "entelechy," an expression that he borrowed from Aristotle while admitting that he changed its meaning. What exactly it means in Aristotle is subject to debate among specialists, so it's better not to offer a translation. Let's just note that, etymologically, it is derivable from the Greek word *telos*, which means goal or end. Now the goal-orientedness of developmental processes was clearly an important aspect for Driesch, but his notion of entelechy contains more than this. I shall come back to his concept in Section 1.3.

Driesch then proceeded to provide a proof according to which the factor *E* cannot be a "machine" (his word for causal mechanism, as we shall see in Section 1.3). As a first premise in his proof, Driesch claimed that, in a harmonious-equipotential system, "each one of its parts behaves like the whole" [my translation] (Driesch, 1905: 207). Thus, if *E* were a machine, it would have to be contained as a whole in all of the parts. As a second premise, Driesch claimed that no machine contains itself in all of its parts. From these premises, it follows deductively that *E* is not a machine. Echoing one of Descartes' proofs for the immaterial nature of the soul from its indivisibility, Driesch claimed that *E* is an immaterial principle or an "intensive" as opposed to an "extensive" term (expressions he borrowed from Kant). Thus, Driesch defended *vitalism*, which is a form of *ontological antireductionism* (see Section 1.2).

Even though Driesch's proof is formally valid, the premises are problematic (Weber, 1999). Nonetheless, the impact of Driesch's reflections on the science of developmental biology was considerable. While the solution he offered, the theory of the immaterial entelechy, never had many followers, it has nonetheless influenced developmental biological thought. For example, Driesch's vitalistic ideas have inspired the concept of morphogenetic field (see Section 2.1) and the theory of positional information (see Section 3.1). Furthermore, it was clear that Driesch had identified a major problem for developmental biology and that the solution was not going to be simple. It should be noted that his argument cannot simply be dismissed by pointing out that all the cells contain a copy of the entire genome, for it still needs to be explained why different cells in an embryo activate different parts of this genome at different times in response to their relative position. We will see what solution was eventually found for this problem in Sections 2 and 3.

This historical episode is very instructive for setting the stage for an engagement with developmental biology from a contemporary philosophy of science perspective. For while vitalism is currently not a live issue (at least in scientific and anglophone philosophical circles), mechanism, reductionism and the extent to which living organisms and their development can be understood like machines very much are.

Since Driesch's days, developmental biologists have learnt a great deal about the processes by which embryonic cells become committed to specific fates within the organism. Long before the molecular turn, experimental embryologists figured out at what stages embryonic cells become determined to form, for example, eyes or neural tissue. By grafting and other experimental techniques, they showed that some cells and tissues interact somehow in this determination process. They also showed in fruit flies that genes are involved. After about 1980, numerous molecules (mostly proteins and the genes that encode them) were identified that mediate such interactions, as well as proteins that control the cells' gene expression patterns such as to commit the embryonic cells to a specific developmental fate. A small number of so-called model organisms were instrumental in these discoveries, including the fruit fly *Drosophila melanogaster*, the African clawed frog *Xenopus laevis*, the zebrafish *Danio rerio*, the mouse *Mus musculus* and the water cress *Arabidopsis thaliana*.

Do these discoveries solve the puzzle raised by Driesch, and if yes, how? Do they constitute a reduction in some sense, or do they vindicate holistic notions such as emergence? What concepts do biologists use when attempting to explain developmental processes? What are the relevant concepts of mechanism and of cause? What is the relationship between the knowledge of classical experimental embryology and that of molecular developmental biology? Did the former provide explanations or was it merely descriptive? Why are molecular accounts deeper than higher-level explanations, if they are? These are the questions to be addressed in this Element. In the rest of this section, I shall outline some philosophical perspectives that could be used for such an undertaking, and eventually choose one.

1.2 Reduction and Emergence

The topic of reductionism has always loomed large over developmental biology, as the debate between developmental mechanics versus vitalism outlined in the Section 1.1 has shown. Driesch opposed a form of what philosophers call *ontological* reductionism or sometimes physicalism, which the early proponents of developmental mechanics endorsed. This is the idea that living matter consists of the same stuff and is subject to the same physical-chemical laws as ordinary matter. Vitalists, by contrast, believed that the vital forces such as Driesch's entelechy could interfere with physical-chemical laws. While most contemporary philosophers of science (and some philosophers of mind) accept ontological reductionism, many believe in addition that this is the *only* viable form of reductionism. Specifically, they reject two other forms of reductionism, widely known (after Ayala, 1974) as *methodological* and *epistemological* reductionism, respectively. The former, which has rarely been defended by philosophers of science, is the idea that all scientific inquiry should use the same approach (e.g., methods appropriate for the lowest level). The latter claims that there is some logical or epistemic relation (e.g., entailment or explanation) between two confirmed bodies of knowledge, one of which is more fundamental than the other. This is the most extensively debated form of reductionism in the philosophy of biology (Brigandt and Love, 2017).

Whatever arguments philosophers have exchanged on this topic, it is surely tempting to view developmental biology's recent spectacular successes in identifying the molecular basis of development as a triumph of some form of reductionism or another. But which form? By rejecting Driesch's vitalism we are only committed to an ontological reductionism. Can we also claim a form of epistemological reductionism?

First, we would need to identify something to be reduced. A part of the philosophical debate on reduction has been about the question of how some scientific theory is related to its successor (e.g., statistical mechanics and classical thermodynamics or wave optics and Maxwell's electromagnetic theory). Is the older theory reducible to the newer, more fundamental theory? This is called diachronic reduction (Nickles, 1973). Or are we speaking about whether some theory that describes a phenomenon at a higher level is reducible to a lower-level theory without these theories standing in a historical succession, which is called synchronic or inter-level reduction? I shall briefly examine these two kinds of reduction with an eye to our topic. As the knowledge of developmental biology is not normally perceived as consisting of theories (Love 2014), I will use the more neutral term "bodies of knowledge."

First, diachronic reduction. The two bodies of knowledge that might be candidates for a historical succession relation are those of classical experimental embryology and contemporary molecular developmental biology. The former was the research program of *Entwicklungsmechanik* already mentioned in Section 1.1. As we have seen, this was a science that experimented with embryos such as amphibians or marine invertebrates in order to study interactions between different parts of the embryo. However, it has been claimed that classical experimental embryology never provided any explanations of the phenomena they studied; the experiments at best described the phenomena to be explained in the first place (Rosenberg, 1997; cf. Laubichler and Wagner 2001). On this view, such explanations only became available once the first molecules involved in development had been identified. This is a strong form of reductionism, which claims that all good explanations are reductive. I will refer to this as *explanatory reductionism*. I will argue in Section 2 that classical experimental biology did provide causal-explanatory knowledge and that there is much diachronic continuity between these two sciences; however, it consisted mainly in the experimental practices.

Second, inter-level reduction. I will focus here on *explanatory reduction*, which encompasses not only relations of a theory to a fundamental theory (theory reduction) but also relations between other knowledge items such as individual facts, generalizations of varying scope, fragments of theories or models of mechanisms (Brigandt and Love, 2017). A central question is the extent in which the properties of complex systems are explainable in terms of the properties of their parts and their organization and interactions. It is crucial to include the organization and interactions in the characterization of explanatory reduction to make this a viable notion. In addition, it is sometimes claimed that explanatory reduction means to appeal to parts that can be studied "in isolation." Kaiser (2015) has argued that this requirement is only viable if understood as "studied in a context other than *in situ*." For example, the standard explanation of the propagation of action potentials (nerve signals) by ion channels located in neuronal membranes is reductive not because these ion channels can be studied in complete isolation (they can't work without a membrane), but because they can be studied in vitro; for example, in small patches of membrane attached to the tip of a pipette.

A major obstacle to explanatory reduction that has been claimed is the existence of so-called *emergent properties* (i.e., such properties of complex systems that are *in principle* unexplainable from the properties of the parts and their interactions). A possible source of this kind of emergence is top-down causation (i.e., an influence of the whole over its parts). A standard way of arguing for top-down causation appeals to situations where what some part of an organism does depends on the activities of the organism as a whole (Dupré, 2021: 3–13). For instance, the movements of the heart valves are influenced by the whole body's physiological state. In addition, it is argued that the heart valves would rapidly decay if it wasn't for the vital activities of the rest of the organism. Thus, it appears that to understand living organisms it's not enough to look *down* to its parts; we always need to look *up* to the whole and beyond, the organism's connections to the world that surrounds it.

It is beyond the scope of this Element to do full justice to such arguments; I only wish to point out here that they typically presuppose that, in attempts at explanatory reduction, what we are seeking are explanations of biological phenomena that are *complete*. When Dupré argues that, for example, his capacity to walk upstairs is not reducible to the capacities of any of his parts because "both the capacities of the parts and their very existence as the kinds of parts they are depend on the whole organism" (2021: 11), he assumes that a successful reduction would have to include everything that is causally relevant to his capacity to walk upstairs, which includes the whole organism and a lot of environmental conditions. But note that this does not preclude a *partial* reductive explanation that correctly identifies *some* lower-level causes or constituent capacities, perhaps even all the salient ones given our explanatory goals. Completeness in the sense of causal sufficiency is not a realistic goal for reductive explanations. According to Kaiser (2015), such explanations only need to appeal to a lower level and satisfy the isolation condition mentioned above. Thus, even if Dupré's metaphysics is correct, suitably understood reductive explanations are still possible.

Of course, that reductive explanations exist doesn't mean that *all* good explanations are reductive. For example, while the standard account of action potentials in neuroscience is reductive (Weber, 2005), most evolutionary explanations are not (Sober, 1999). We can accept some reductive explanations or explanatory reductions without committing to explanatory reduction*ism*. As we shall see, in developmental biology we find explanations of the reductive and of the non-reductive type. Classical experimental embryology provided non-reductive explanations, while the explanations of molecular developmental biology are typically reductive (but incomplete).

As we shall see in Sections 2 and 3, it is indeed the case that developmental biology succeeds by identifying only a small fraction of all the causally relevant factors that are present in a developing organism and by ignoring a vast range of other factors. Explanations in biology, whether they are reductive or not, often involve *abstractions*. This means that they leave out a lot of causal detail and focus on just some select causes that are deemed pivotal with respect to the goals of inquiry. Large parts of the organism and its environment are backgrounded. Furthermore, scientific explanations often represent causal relations in an *idealized* way, that is, by radically simplifying the way in which causes operate (see the example of morphogens in Section 3). Such simplification is not a defect. First, according to some philosophers, it can provide *understanding* (Potochnik 2017). Second, it has tremendous *heuristic* value for research (Wimsatt, 2007: Chapter 6). Once it is understood that biology doesn't aim at complete explanations, I suggest, notions such as top-down causality and emergence are revealed as being irrelevant to scientific practice. Of course, it is still possible that some phenomena resist even partial reductive explanations. The phenomena discovered by Driesch (see Section 1.1) are a potential candidate, and I shall examine them in Section 3.4.

It will not be possible to give the topic of reduction a full treatment here. Nonetheless, I will point to some potential difficulties for explanatory reduction (see Section 2.5), even though I think there clearly are such explanations, however partial, in developmental biology.

1.3 Mechanism

As we have seen in Section 1.1, a central issue raised by Driesch and *Entwicklungsmechanik* has to do with mechanism or the doctrine that all biological phenomena can be explained mechanistically. But what exactly should we take this to mean? To start with, it is important to distinguish between (1) mechanism as the doctrine according to which life can be understood mechanistically, which should really be called "mechanicism," (2), the idea of a mechanism as a machine-like structure and (3) the notion of causal mechanism (Nicholson 2012). While all machine-like structures are causal mechanisms, the converse may not hold; for example, molecular diffusion is a causal mechanism, but it is not machine-like (Levy 2014). As we have seen in Section 1.1, Driesch's proofs primarily purported to show that no machine-like structure that is composed of different parts can be responsible for the phenomena of regulative epigenesis; but I will show now that his concept of machine was so broad that we must read him as being opposed to any kind of causal mechanism that is consistent with the laws of physics and chemistry.

Driesch's concept of a machine was that of an "extensive manifold" in contrast to entelechy, which he viewed as an "intensive manifold." "Extensive" here can be read in the Cartesian sense of spatially extended. Driesch characterizes extensive manifolds as a form of "causality that is based on spatial configurations" (Driesch, 1928: 142) and as a "physical-chemical structure" or a "tectonic," which contains "numerous physical and chemical substances and forces in a typical order" (Driesch, 1905: 206 [my translations]). It is not entirely clear what Driesch meant by "physical-chemical structure" and by a "tectonic," but the most natural reading would be that he meant any kind of causal mechanism that posits different physical or chemical causes in some spatiotemporal arrangement that act in accordance with physical-chemical laws, not just such mechanisms that resemble a human-made machine. Indeed, Driesch's characterization sounds a bit like contemporary accounts of what causal mechanisms are.[2] That he wanted to exclude any kind of causal mechanism is also implied by the fact he thought that the only feasible alternative is an immaterial force.

But what is a causal mechanism? And what is a mechanistic explanation? Even though these questions are related, the second one may be easier to answer. We can distinguish between at least two senses of "mechanistic explanation": first, the explanation of an activity of a system that breaks this activity down into the activities of components and shows how the interactions between these components produce the system's activities. For example, the activity of neural cell membranes of transmitting action potentials is broken down into the opening and closing of selective ion channels located in the cell membrane. Such mechanistic explanations involve some system (here: a neural cell membrane) with an activity (here: transmitting action potentials) and a set of components (here: selective sodium and potassium ion channels) with their own activities (selective ion transport, voltage-dependent opening and closing) that together produce the activity of the system. Some philosophers of science known as "New Mechanists" think that scientific explanations essentially describe such mechanisms.[3]

According to an alternative and perhaps also more common conception, a mechanistic explanation is simply a causal explanation that identifies some *mediating causal variables* for a cause–effect relation. Simply put, a mechanistic explanation in this sense is a causal explanation that shows how some cause–effect relation is mediated by causal variables that lie causally in between the cause and the effect. For example, when it is shown that smoking causes lung cancer due to certain carcinogens damaging the DNA of lung cells, this amounts to a mechanistic explanation in the second sense. Smoking causes the release of carcinogens into the airways and their uptake by lung cells, which causes DNA damage in the cells' nucleus, which destroys some of the cells' systems that control their division, which causes uncontrolled cell division (i.e., cancer). Of course, such mechanistic explanations may involve not only linear causal chains but also more complex causal networks including feedback and dynamics. This alternative conception (e.g., Baetu, 2019) is distinguished from the New Mechanism approach by its non-verticality, that is, by its not referring to distinct levels of organization. Both kinds of mechanistic explanation may be found in developmental biology (Baedke, 2020).

---

[2] For example, "Mechanisms are entities and activities organized such that they are productive of regular changes from start or set-up to finish or termination conditions" (Machamer et al., 2000: 3).

[3] Craver (2007) is still the most elaborate articulation and defense of this view.

The two notions of mechanism also have implications for experimental methodology. For according to New Mechanism, the components of a mechanism and their activities do not *cause* the phenomenon to be explained; they rather *constitute* it. To test for such constitutive relations, Craver (2007) proposed the criterion of mutual manipulability or MM. Roughly, MM is the idea that some component (e.g., a molecule) belongs to a mechanism for some phenomenon if (i) an experimental "bottom-up" intervention on that component changes the phenomenon and (ii) a "top-down" intervention on the phenomenon brings about a change in the component. Clause (ii) is meant to exclude factors that affect the operation of a mechanism but are causally too remote to belong to the mechanism (such as the effect of blood sugar levels on cognitive activities in the brain). MM has been criticized especially for the idea of top-down interventions, which may not be possible in principle because a system and its parts cannot be manipulated independently (Baumgartner and Casini, 2017). In response, New Mechanists (Craver et al., 2021) have recently revised the mutual manipulability condition in a way that may move their account closer to a level-independent mediating-variable account of mechanism. Thus, the two different conceptions of mechanistic explanation may turn out not to be so different in the end.

Other critics of New Mechanism have presented various examples of scientific explanations that do not appear to be mechanistic, for example, natural selection explanations in evolutionary biology (Skipper and Millstein, 2005) or systems-biological explanations (see the essays by Mekios and Gross in Braillard and Malaterre, 2015). Especially explanations that use dynamical equations are thought to represent a completely different kind of explanation (Stepp et al., 2011). However, this may just be too narrow an understanding of mechanistic explanation (Kaplan and Bechtel, 2011). Silberstein and Chemero (2013) argue that there are neurological systems that exhibit such a high degree of interaction that they cannot be decomposed and localized into separately operating parts. To the extent that mechanistic explanation requires such delocalization and decomposition, such systems are not mechanistically explainable.

Other critics have attempted to show that there exists an important class of biological explanations that cite *pathways*, which differ from mechanisms in several respects (Ross, 2020). One major difference is that pathways track the flow of some specific entity (e.g., a metabolite) through a series of steps without paying attention to much of the other causal factors that are necessary for these steps to occur. According to such critics, it is not illuminating to assimilate concepts such as the pathway concept to the mechanistic framework because by doing so we lose sight of the diversity of explanations that exist in biology.

I shall examine a specific kind of pathway, namely signaling pathways, in Section 4.2.

Another problem for New Mechanism, according to some critics, is that mechanistic accounts of biological systems usually assume a fixed inventory of entities, while especially in developmental processes, entities and constitutive relations come and go (Mc Manus, 2012; Parkkinen, 2014). An example is the spindle apparatus that cells need in order to pull the previously duplicated chromosomes apart during cell divisions, which is assembled anew before every cell division. During embryonic development as well, there are transient entities such as the Spemann–Mangold organizer (see Section 2) that form and disappear again. What is more, Dupré (2013) criticizes that standard mechanistic accounts assume that mechanisms are made up of stable things. He objects that "the entities that form the hierarchy of biological ontology are not stable. They are, rather, stabilized over a wide variety of timescales, and the processes of stabilization are a fundamental part of the explanation of the activities of living systems" (30). According to Dupré, a process perspective (see Section 1.4) is better fitted to living systems than any mechanistic perspective. While New Mechanists, of course, do not deny the importance of processes, they tend to downplay the extent in which mechanisms are merely abstractions from much more complex processes.

Proponents of New Mechanism are typically not impressed by such objections, and perhaps rightly so. For they are disposed to apply the concept of mechanism quite generously. For example, they will hardly feel any pressure to conceive of mechanisms as being rigid in terms of the entities that constitute them. Why not have mechanisms that assemble some of their own constitutive parts during their operation? And what's wrong with a mechanism that loses some parts while it plays out? New Mechanists don't think that all mechanisms are basically like a Swiss watch, which is complex but rather rigid. Developmental mechanisms are much more dynamic entities (Baedke, 2021).

While there is much that we can learn from New Mechanism-style analyses, there are also some aspects which are particularly relevant to developmental biology that this approach will fail to illuminate. One such aspect is related to the issue of *causal selection*, which I will turn to in Section 1.5, after briefly reviewing some other perspectives on developmental biology.

1.4 Other Philosophical Perspectives on Developmental Biology

There are many philosophical perspectives that can potentially illuminate the practice of developmental biology, in addition to the ones that we have just discussed in the previous two sections (Love, 2020a). These perspectives differ considerably both in the approach that they take and in their aims. Some are more metaphysical, others more epistemic. I just would like to mention four more perspectives.

First, philosophers and biologists have debated what exactly development is. Does it stretch over the whole life of an organism or does it end sometime before death? Different conceptions of development lead to different answers to this and related questions (Pradeu et al. 2011).

Second, there is the perspective of process philosophy (Nicholson and Dupré, 2018). This approach focuses on the ontology ("the science of what there is") of living organisms and their parts and takes them not merely to *contain* processes but essentially to *be* processes. On this view, any living thing such as a complex animal, a plant or a single cell isn't really a "thing," that is, a three-dimensional object moving through time; it is rather a process that is extended in time. Thus, processes are not owned by things that exist independently of them; rather, things *are* stabilized processes. Of course, processes may interact, and they can be decomposed into parts, which are themselves processes. Mechanisms are also processes (or abstractions thereof). It should be clear that process philosophy is an attractive option for developmental biology, which basically studies processes. For example, a process perspective has proven to be helpful when it comes to better understand the criteria by which developmental biologists identify organismic life cycles and developmental stages (see the essay by James DiFrisco in Nicholson and Dupré 2018).

Third, there is developmental systems theory or "DST." This approach started by taking issue with dichotomous accounts of development according to which some parts of an organism, most frequently its genes, are considered as harboring genetic "information," a "program," "blueprint" or "instructions" while all the myriads of other parts that make up a developing organism merely execute. According to DST, it is the entire developmental system or "develop-mental matrix" that builds an organism, and the differences in causal role between genes and other parts (which are not being denied) do not justify the abovementioned distinctions (Griffiths and Gray, 1994; Oyama, 2000). Debate has focused in particular on the idea of a "causal parity" between genes and other developmental resources (Griffiths et al., 2015; Waters, 2007; Weber, 2006; forthcoming-a). This debate has been productive, as it has led to several potentially fruitful ways of distinguishing between different kinds of causal relations. While I shall not enter into the causal parity debate in this Element, some of the concepts and distinctions that came out of this debate will prove to be helpful for my causal approach, to be outlined in Section 1.5.

Fourth, there are several more practice-oriented perspectives. For example, some researchers in history and philosophy of biology have analyzed in particular the use of experimental systems and model organisms that play such a central role in developmental biology (e.g., De Chadarevian, 1998; Weber, 2005). Another practice-oriented perspective is the erotetic approach (Love, 2014), which looks at the way in which research in developmental biology is organized around research questions rather than theories.

All the perspectives mentioned in this section have been helpful in order to understand some aspects of developmental biology, and we should resist the temptation to consider any one of them as being able to account for everything that goes on in science. If I have chosen to apply a particular philosophical perspective in this Element, namely a *causal* perspective (see the Section 1.5), it is not because I take this perspective to be fundamental. In fact, the term "perspective" is deliberately chosen here to indicate that we can apply different conceptual and philosophical frameworks to scientific practice, depending on the questions that we are seeking answers for. It is doubtful that any single framework will give us anything close to a complete picture of scientific practice nor of the reality that it studies. Nonetheless, I contend that there are some key concepts and practices in developmental biology that can be best understood by taking a *causal* perspective, drawing on some results coming from the philosophical study of causality, in particular attempts of distinguishing between different kinds of causes, but also on the role of abstraction and idealization in causal reasoning. Furthermore, I will show that focusing on causality goes some way toward explaining developmental biology's epistemic success, both before and after the molecular turn.

In the following section, I explain what it means to take a causal perspective.

1.5 Taking a Causal Perspective in the Analysis of the Practices of Developmental Biology

In taking a causal perspective, I focus on the biologists' search for causes of developmental change by means of experimentation as well as modeling, with special attention to the *kinds* of causes that they are seeking. This characterization of the causal perspective requires some explication. I will begin this explication by first outlining the basic account of causality and causal explanation that I shall be presupposing throughout the Element, namely an *interventionist* account. Then, I will explain what I mean by "kinds of causes," and what the use is of distinguishing between different kinds of causes. Finally, I will explain what the difference is between my causal perspective and the philosophical approach known as New Mechanism (see Section 1.3). I want to be clear from the outset that my causal perspective is also in some sense a mechanistic perspective – Driesch would certainly recognize and reject it as such – but it is distinct from New Mechanism in significant ways.

An important method for identifying causes involves experiments, and indeed developmental biology has been a thoroughly experimental science, at least since the days of *Entwicklungsmechanik*. Contemporary philosophical accounts of causality acknowledge the close link between causality and experiment,[4] and they even define causes in terms of idealized experiments. One of the most influential accounts of this kind has been developed by Woodward (2003). In a nutshell, he defines an idealized experiment as an intervention that, like a good surgeon, precisely targets a causal variable.[5] Then, some variable C is a cause of a variable E exactly if C can be used to manipulate E by idealized experiments in this sense. Of course, idealized experiments can rarely be realized in biology, but the notion is useful both as an epistemic ideal and for the purpose of defining causality.[6]

According to Woodward, manipulability is also the key to scientific explanation. On his view, we have "at least the beginnings of an explanation when we have identified factors or conditions such that manipulations or changes in those factors or conditions will produce changes in the outcome being explained" (Woodward, 2003: 10). Woodward cites the biologist Robert Weinberg as drawing a contrast between descriptive and explanatory biology and as identifying the latter pretty much with molecular biology. I think this identification is a mistake. According to Woodward's manipulability account of explanation, which I will adopt in this Element, molecular biology's explanatory power is due to the fact that molecular techniques greatly increase the experimenters' control over the processes they study. Thus, it is not because molecules or the physical chemistry of life are fundamental that molecular biology offers deeper explanations of some biological phenomena than its predecessors. Where molecular explanations are deeper than higher-level explanations, which isn't

---

[4] I am not claiming that uncovering causal relations is the only purpose of experiments; see Weber (2005: Chapter 6) for some other uses of experiment in biology.

[5] Somewhat more precisely, an idealized experiment is an intervention that sets a variable C to a determinate value and thereby changes the value of another variable E without changing the values of any other variables that are causes of E except those that lie on a path between C and E.

[6] The reader might also note that the idea of an intervention is itself causal and wonder if this doesn't make the account circular. According to Woodward, it is circular but not viciously so because as agents we have an intuitive idea of what an intervention is and we can use this idea in order to clarify an idea that is less intuitive, namely the idea of cause. However, unlike in earlier versions of interventionism, causality is not reduced to human agency. The account is nonreductive, i.e., it does not aspire to define causality from scratch, that is, from purely noncausal notions. What it does achieve is to exhibit important conceptual links, such as the ones between the notions of ideal experimental intervention and various causal concepts.

always the case (Sober 1999), it is because molecular biology affords more *control* over the entities and behaviors in its domain of inquiry than pre-molecular biology. However, as we will see in Section 2, it is wrongheaded to think that developmental biology was merely descriptive before its molecular turn in the 1980s. Experimental embryologists did identify some causal variables that afford control over developmental processes, only less than molecules.

In my initial characterization of the causal perspective, I mentioned different *kinds* of causes. This idea, too, requires some elaboration. Most philosophers of causality – but not Woodward – have accepted a thesis that is known as "Millean parity" (after the philosopher John Stuart Mill). This thesis concerns our widespread practice of singling out individual causal factors from an entire field of factors that are necessary to bring about an effect. For example, we may blame a pyrotechnic device or an electric spark for causing a wildfire even though oxygen and the presence of dry organic substances were equally causally relevant for the fire. We call the firework or the spark "the" cause, while any other necessary causal factor is merely "a" cause or backgrounded altogether. Millean parity is the claim that such a discrimination has no basis *in* the causal relations in question. At best, it reflects what we happen to be interested in; perhaps we tend to focus on those causes on which we could have intervened to prevent an accident.

More recently, philosophers of science have started to resist Millean parity in the context of scientific explanations (Baxter, 2019; Franklin-Hall, 2015; Lean, 2019; Plutynski, 2018; Ross, 2018; Waters, 2007; Weber, forthcoming-a; Woodward, 2010). The starting point for such attempts is the realization that not all causal relationships are alike; they may differ in various respects. Here are some such differences:

(1) Some causal links are more *stable* than others, that is, they hold under a greater range of background conditions (Woodward 2010).
(2) Some causes are *proportional* with respect to their effect (Woodward 2010). This means that the concepts used to pick out the causes and effects have the right level of generality. For example, if trained pigeons peck at all red spots, to say that they peck at scarlet spots is true but not proportional.
(3) Some causal relations are *specific* in the sense that one type of cause has only one type of effect (Woodward, 2010), or where the cause exhibits some kind of preference for one (or a few) of its effects. While it might be difficult to say in general what it means to say that a cause shows a "preference" for one of its effects, in the realm of molecular interactions there is a clear sense in which this can be the case: some molecules bind

other molecules with a higher affinity than others (Lean, 2019). What this means for a ligand L and its binding partner P is that it takes a lower concentration of L to keep half of the Ps bound (due to thermal motion, there is in molecular interactions typically a dynamic equilibrium between association and dissociation of two binding partners). *Binding-specific causes* in this sense are extremely important in developmental biology, as we shall see.

(4) Some causes are more specific than others in an altogether different sense (Griffiths et al., 2015; Waters 2007; Weber, 2006, forthcoming-a; Woodward 2010). Some philosophers mean by this that some causes afford *fine-grained control* over their effects, like a light dimmer as opposed to a mere on-off switch. As we shall see in Section 2, this sense of causal specificity also plays a role in developmental biologists' attempts to understand highly complex systems.

(5) Some causes are *actually* variable in a population and fully or partially account for actual variation in the effect variable, while others are not actually variable and make a difference only *potentially* (Waters 2007). We shall see that this distinction is also useful in developmental biology, but only as an idealization (Section 3.2).

I will show in Section 4 that developmental biology has also come up with distinctions within causality of its own, in particular the distinction between instructive and permissive causes. I will also introduce a new property that only some causal relations have, namely causal coherence. For now, it is sufficient to note that a cause isn't just a cause; there are numerous ways in which they can differ.

Distinctions between causal relationships are significant because they can help us understand the scientific practice of picking out in scientific models or explanations those causal factors that are particularly useful to know for achieving certain epistemic goals relevant to biological systems. This practice is called *causal selection*. It should not be confused with the general practice of causal *inference*, that is, the problem of demonstrating the existence of causal relationships with the help of experimental (or observational) data, although in some cases the same data may be used for demonstrating the existence of a causal relationship and for establishing the features that make the cause explanatorily salient.

The claim that not all causally relevant factors are equally explanatorily salient in developmental biology can be illustrated with the following examples: the fact that cyanide leads to an arrest of developmental processes is not so interesting for the developmental biologist because cyanide is known to poison the cell's energy metabolism and it is pretty trivial that developmental processes require energy to go forward. By contrast, knowing a set of molecules that have been demonstrated to pattern the embryo along one of its axes – if there are indeed such molecules; see Section 3 – seems highly explanatory salient in developmental biology. Causal selection is particularly relevant where thousands of causal factors, which may be found at different scales or levels, contribute to a specific phenomenon of interest. What is more, biological reality contains a multiplicity of different structures, but no *general* structure that, if known, would allow biological science to explain all the phenomena in its domain (Waters 2017). For this reason, biologists must look for things that have some special features other than generality.

Many biological explanations *abstract* from many of the myriads of details that make up biological reality and isolate a few factors as explanatorily pivotal. The practice of causal selection can use such distinctions between causal relations as I have just mentioned them as guides to this kind of abstraction. One strategy consists in focusing on causes that *actually* make a difference in a population of individuals (Waters 2007). Another puts the spotlight on *specific* causes (e.g., in the one-to-one sense; see Lean, 2019), or such causes that we can use to *control* some outcome, including fine-grained control (Ross, forthcoming). In some explanatory contexts, scientists will be looking for causes that can bring about some effect under a variety of background conditions (i.e., they will use *stability* as a guide to causal selection). The principles of causal selection used are not always the same; they depend on the investigative context. We shall see in subsequent sections what strategies of causal selection developmental biologists use.

Biologists not only *abstract* often from causal details, sometimes they also work with *idealized* causal models. A common way of distinguishing between abstraction and idealization is that the former only leaves out details, while the latter deliberately introduces outright falsehoods into scientific representations. Many scientific models make deliberate counterfactual assumptions such as infinite population size, frictionless motion, or point particles. Some philosophers think that the main purpose of such idealizations is to build representations that resemble reality to some degree (Weisberg, 2013). Others believe that false models can provide direct understanding about the world (Potochnik, 2017). I side more with Wimsatt (2007: Chapter 6) who sees idealizations mainly as a heuristic for guiding research. As I shall show in Section 3, developmental biology uses idealized models in order to learn more about causal relations in developing organisms. As we shall see, some of the distinctions within causality mentioned earlier turn out to be idealizations themselves, and this is in part why they can be useful for biologists.

Finally, I will now explain how my causal perspective differs from New Mechanism (discussed in Section 1.3). Before I begin, here are some common features: it must be acknowledged that New Mechanists also recognize causal or "etiological" explanations, which they contrast with their "constitutive" explanations. The former explain by exhibiting a phenomenon's causal history, while the latter show how upper-level phenomena are constituted by lower-level level components and their interactions (Craver 2007). Furthermore, it should be noted that issues related to causal selection arise also within the framework of New Mechanism. Mechanistic explanation requires the selection of factors that belong to a mechanism and factors that don't. New Mechanists have sought to solve this problem with the criterion of mutual manipulability; however, this proposal has remained controversial (see Baumgartner and Casini 2017). Furthermore, it doesn't apply to purely etiological explanations.

Now for the differences. The first difference concerns the commitment to a hierarchy of *levels* that defines some conceptions of mechanism found in the philosophical literature. While levels do play a role in developmental biology (Baedke, 2021), not all developmental explanations involve mechanistic hierarchies. The examples I study in this Element mainly involve causes that have an effect later in development and would therefore be considered as etiological by New Mechanists. Thus, the hierarchical or vertical aspects will not enter into the picture. This means that we will be dealing mainly with mechanistic explanations in the mediating variables sense (see Section 1.3), some of which will involve complex dynamics.

The second difference concerns the dominant criterion for explanatory force. For New Mechanists, an explanation necessarily requires knowledge of mechanisms, that is, knowledge of entities and activities that produce a phenomenon. While some leading New Mechanists accept that mechanistic explanations may involve abstraction, they insist on a minimal standard of *completeness* relative to what is to be explained (Craver and Kaplan, 2020).

By contrast, on the account of explanation that I favor, there is no minimal standard of completeness. Of course, to know more mechanistic details will often increase the depth of an explanation, but we can have bona fide causal explanations lacking details that New Mechanists find crucial (Weber, 2008). Classical experimental embryology, too, will come out as merely descriptive according to New Mechanism, which turns out to have a strong reductionistic bent (Rosenberg, 2020). This doesn't do justice to the rich store of causal knowledge that developmental biologists built long before the first molecules were implicated in development.

In any case, on my view the amount of mechanistic detail is not the only relevant consideration. All causally relevant factors are not equally explanatory, as our earlier discussion of Millean parity and causal selection shows. Philosophers of causality have thought hard about how to distinguish causes from mere correlations; however, we still lack a general account of what makes some causal factors explanatorily salient or more salient than others (Woodward and Ross, 2021). This is not the place to develop such an account, but I want to show in this Element that studying the practice of developmental biology reveals some interesting patterns in this respect.

1.6 Focus and Outline of the Element

I will focus in this Element on research about some early events in animal embryogenesis, prior to and during gastrulation in vertebrates or segmentation in the insect larva. During these stages, the animal's basic body plan is laid down, and many cells become committed to specific developmental paths. For example, in vertebrates the cells are differentiated at these stages into the three germ layers endoderm, mesoderm and ectoderm. The endoderm gives rise to the gastrointestinal tract and some of the internal organs such as the liver or the lungs. The mesoderm forms bones, muscles, cartilage and the circulatory and lymphatic systems. The ectoderm forms the nervous system and the skin (with teeth, hair, nails, etc.). Some cells already acquire a specific fate such as neural tissue at the gastrula stage.

The guiding research question in this area of developmental biology is how a group of cells, which have exactly the same genes and do not initially differ in their potential (unlike what Roux thought; see Section 1.1), are determined to follow different pathways. This determination is accomplished by different cells activating and deactivating different sets of genes. Thus, the question is what causes these different cells to activate and deactivate different genes. What is particularly striking in these events is the remarkable *robustness* and *scale invariance* of these processes. This means that the early embryo is often able to produce a normal form even when parts are removed and independently of its absolute size. These phenomena have puzzled developmental biologists ever since Driesch's groundbreaking experiments (Section 1.1). Even though it has long been speculated that these phenomena have something to do with so-called morphogen gradients, satisfactory answers to these questions have emerged only very recently.

Much recent research in philosophy of science has focused on explanation, but I believe we must also pay attention to the way in which causal knowledge can be used for learning even more about biological processes. A powerful example of this is provided by classical genetics. As Waters (2008) has shown, classical genetics provided biologists with much more than just explanations for certain regularities of gene transmission; it also provided powerful investigative strategies based on identifying mutants, genetically analyzing them and recombining them in order to learn more about biological processes. This approach has been extremely important in developmental biology; in fact, none of the molecular discoveries that I shall discuss in subsequent sections would have been possible without it. I will show here that other kinds of causal knowledge, in particular knowledge from classical experimental embryology, have been similarly important (see Sections 2.3 and 2.4).

It must be admitted that in taking a causal perspective we are also going to miss many fascinating aspects of the practice of developmental biology. This science in particular is characterized by very rich *descriptions*, e.g., of anatomical structures of embryos or of the series of changes that a developing embryo undergoes. The process of development is typically divided up into stages, such as fertilization, cleavage and formation of the blastula (a hollow ball of cells), gastrulation and specification of the germ layers, neurulation, organogenesis and limb formation (these are the typical vertebrate stages; of course, insects are different). These practices, too, involve various kinds of idealization (Love 2020a). The descriptive practices of developmental biology alone could be the subject of a book-length philosophical study. We are also going to miss out on important practices such as tracing the lineages of cells (i.e., determining which embryonic cells give raise to which structures later in the embryo). In focusing on causality, I do not mean to imply that such practices are irrelevant to the success of developmental biology.

What I hope to show is that by attending to causality and to different kinds of causal relations we can better understand the nature and significance of some key concepts used in developmental biology such as induction (Section 2.1), organizer (Section 2.3), morphogen (Section 3), instructive versus permissive cause (Section 4.1), signaling pathway (Section 4.2) or selector gene (Section 4.3). Interestingly, most of these concepts (except signaling pathway) predate the molecular turn. Furthermore, we can also identify some experimental and reasoning strategies that can, to a large extent, account for the remarkable *success* of developmental biology in the last 100 years or so. This progress is in part a result of a tremendous growth of possibilities to experimentally *intervene* on developmental processes in order to learn more about them. I am not claiming that all explanation is causal, nor that experimentation is the only method that contributed to developmental biology's success. As we shall see, theorizing and mathematical modeling have also played a role. But I contend that different forms of causal reasoning, including experimental, theoretical and model-based forms, were a main driver of progress, because many causal factors identified proved to be invaluable tools for further research. Explanatory considerations should not be separated from this context of inquiry.

Without further ado, let's see the causal perspective at work in trying to make sense of some explanatory and discovery practices of developmental biology. Section 2 will focus on the practice of classical experimental embryology and its transition to molecular developmental biology. In Section 3, I will examine so-called "morphogens" and the role of an idealized causal model concerning their action. In Section 4, I will provide further analysis of some causal concepts that are operative in developmental biological research practice. Finally, in Section 5 I draw some general conclusions about the nature of scientific progress in such a thoroughly experimental science as developmental biology.

2 Useful Causes: The Quest for Inducers and Organizers

In this section, I will first examine two concepts that were introduced during the classical period of experimental embryology, which are still in use today, namely the notions of *embryonic induction* and *morphogenetic field* (Section 2.1). Then, I will present a classical experiment involving embryonic induction, namely the famous Spemann–Mangold experiment. I will also discuss some later findings that challenged the conclusions drawn by Spemann and Mangold from their experiment (Section 2.2). In Section 2.3, I will present a causal analysis and a proposal as to wherein the scientific significance of this experiment consisted. In Section 2.4, I will argue that the causal knowledge that was handed down from classical experimental embryology, while being explanatory in its own right, had considerable heuristic value for identifying some of the molecules mediating some of the organizer's effects. Finally, Section 2.5 will examine the question of whether these molecular findings should be viewed as reductions or rather as a replacement of the classical body of knowledge.

2.1 Classical Concepts: Embryonic Induction and Morphogenetic Fields

Early on in the twentieth century, classical experimental embryology established the importance of interactions within the embryo or "epigenesis" in committing embryonic cells to a specific fate, at least in vertebrates (see Section 1.1). Embryologists were keen on learning more about these interactions and began to devise sophisticated experiments, using for the most part amphibians as experimental organisms. New questions arose, such as the following: At what developmental stages are different embryonic cells committed to their fate; for example, epidermis (skin), neuroectoderm (brain and nerve fibers), mesoderm (muscle and bones, blood vessels) or endoderm (gut, internal organs)? What interactions determine this fate? And what are the chemical mediators of these interactions, if they are chemical in nature? Two classical concepts that emerged in pursuit of such questions were those of *embryonic induction* and *morphogenetic field*.

The notion of induction was introduced by Hans Spemann as a result of his experiments on eye development in Northern European newts of the genus *Triturus*. Spemann found in 1901 that newt embryos failed to develop a lens when he destroyed the eye rudiment underlying the epidermis that normally forms a lens. This suggested that the eye rudiment or optic vesicle (an outgrowth of the early brain) somehow caused the epidermis cells to change fate and form lens tissue. Indeed, W. H. Lewis showed in 1904 that optic vesicles transplanted into the flanks of frogs induced ectopic lenses ("ectopic" means that the structure is in a place where it doesn't belong). Lens induction thus became a sort of a paradigm for embryonic induction (Saha, 1991). However, this simple idea was soon overthrown by various recalcitrant findings and objections.

For one, it was found that lenses can sometimes form in the absence of optic vesicles, a phenomenon that came to be known as "free lenses." Furthermore, Lewis's purported proof of sufficiency was eventually rejected because it was possible (and indeed the case) that the transplanted tissue was contaminated by epidermis cells that were already committed to form lenses. Eventually, the simple model of lens induction had to be replaced by a more complex, stepwise process where some of the steps require multiple feedbacks from the induced to the inducing tissue. Nonetheless, Spemann (1936: 26) rightly insisted that it was an *interaction* between neural tissue and epidermis that explains why the lens conveniently forms exactly above the optic vesicle during eye development, and not the unfolding of some preexisting harmony involving only self-differentiation (cf. Roux's mosaic theory mentioned in Section 1.1). This turned out to be an important piece of knowledge withstanding the test of time.

The phenomenon of induction, also referred to as "evocation" by some authors, became a central topic of experimental embryology. In C. H. Waddington's formulation, "[t]wo neighboring parts of an egg or embryo may react with one another, in such a way as to change the capacity for development of one, or perhaps sometimes both, of the reactants" (Waddington, 1956: 16). Waddington also emphasizes that "[b]y interactions between parts which have newly come together, the composition of the embryo gradually increases in complexity" (17). The primary example of evocation mentioned by Waddington is the induction of neural ectoderm by the mesoderm from a region known as the "gray crescent" in amphibian embryos, which contains the Spemann–Mangold organizer (see the Section 2.2). It should be noted that the reciprocity of induction (i.e., the fact that in some cases development will proceed only if the induced tissue gives feedback to the inducing tissue) is already integrated into Waddington's characterization of evocation.

Later, a distinction was introduced between "instructive" and "permissive" inductions, where the former corresponds to Waddington's notion of evocation. I shall discuss this distinction in detail in Section 4.1. For now, it suffices to note that induction is a classical causal concept from experimental embryology that is still being used today, and that induction has long been considered to be a complex multistep process involving feedback from the responding tissue.

Another notion that was handed down from classical experimental embryology is that of *morphogenetic field* (Davidson, 1993; De Robertis, 2006). It is often attributed to Alexander Gurwitsch (1922), who took his main inspiration from Driesch's notion of a harmonious-equipotential system (see Section 1.1). Gurwitsch drew an analogy between some embryonic regions and physical force fields, in his words a "spatial region in which by the specification of the coordinates of an arbitrary point the entirety of the influences on an object located at this point is uniquely determined" (392). In embryology, the "influences" in question are tissue or cell interactions that determine a region's fate in subsequent development. According to another characterization, given by Huxley and De Beer (1934: 276), a morphogenetic field is a "region throughout which some agency is at work in a co-ordinated way, resulting in the establishment of an equilibrium within the area of that field." Unlike Driesch and Gurwitsch with their holistic entelechies and *Ganzheitsfaktoren*, the English authors already conjectured that some gradient in metabolic activity would be able to provide some kind of stable equilibrium. This turned out to be a remarkable prediction (see Section 3.4).

Waddington (1956: 17–18) saw a close connection between the notions of evocation (as he understood it) and morphogenetic field. In his view, any kind of induction requires a morphogenetic field because the inducing and responding tissues must somehow form an integrated whole.

If induction is clearly a causal concept, the notion of morphogenetic field seems richer. While it is often introduced with the help of physical analogies – in particular, force fields in electrodynamics – most classic authors have emphasized the "coordinated" or "integrated" character of embryonic fields (see Huxley and De Beer, 1934: 276 or Waddington, 1956: 25). It is as if parts of the embryo "know" their position and size with respect to the whole system and are capable of adapting their development correspondingly. Of course, there is no real cognition involved; but it is notoriously difficult to explicate notions such as "coordination" or "integration" in purely causal terms. I contend that such notions are psychomorphic metaphors standing for complex causal relations that require more analysis. I will not suggest that embryonic development is actually psychomorphic, but it is likely that psychomorphic notions have been heuristically useful.[7] In fact, in what follows I want to emphasize in particular the heuristic usefulness of classical embryonic notions, not only for theorizing but also, and in particular, in experimental practice.

2.2 The Spemann–Mangold Experiment and Its Discontents

Probably the most famous experiment in classical embryology was performed just about a century ago by Spemann's Ph.D. student Hilde Mangold. She cut a piece of embryonic tissue from the so-called blastopore lip located at the dorsal (back) side of newt gastrulae and grafted it to the bellies of embryos from another, closely related species. The gastrula is a vertebrate embryo that has undergone gastrulation, a process during which a part of the blastula – a hollow ball made up of one cell layer – invaginates and moves inside the embryo, thus forming a multilayered structure. This structure is already differentiated into the three germ layers: endoderm, mesoderm and ectoderm. The blastopore lip lies in the gray crescent area in amphibian embryos, which is where the invagination starts at the beginning of gastrulation. Mangold and Spemann observed the formation of a complete secondary embryonic body axis, as witnessed by a second notochord as well as head and tail and additional embryonic structures in the recipients of the graft (a version of the experiment can be seen in Figure 1). Because this newly formed tissue showed the pigmentation of the recipient and not that of the donor species, Mangold and Spemann concluded that the graft had organized the foreign material of the recipient tissue into a secondary embryo rather than just growing into a secondary embryo itself. Therefore, they introduced the concept of "organizer" or "organization center" to designate the transplanted area that had this power. This was widely perceived as a major scientific breakthrough, earning Spemann the Nobel Prize in Physiology or Medicine in 1935. (Hilde Mangold did not share the Nobel because she died in an accident just a few months after the publication of her famous experiment.)

---

[7] In the closing paragraph of his 1936 book, Spemann (1936: 278) suggests that field phenomena, in particular, bear a strong resemblance to psychological processes and that this resemblance puts us psychic beings at an advantage to understand these processes. For Driesch, of course, this was more than just a resemblance; he saw embryonic development, human cognition and purposeful action as different manifestations of the same fundamental force of nature, namely what he called *entelechy*.

(a)

(b)

(c)

(d)

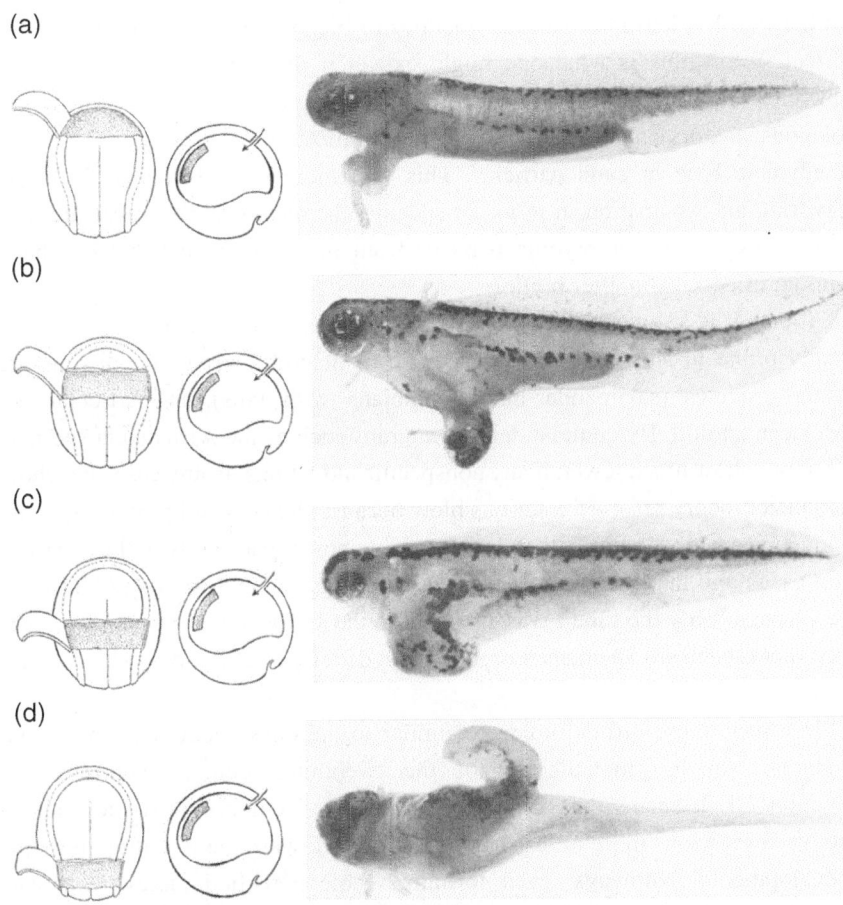

Figure 1 Regional specificity of inductive power by the archenteron roof transplanted into the cavity of newt gastrulae (image reproduced from Mangold, 1933). This is a variant of the classic Spemann–Mangold experiment where dorsal tissue is inserted into the cavity of an earlier embryo. It is called the *Einsteck* method after the German word for insertion. The drawings in (a)–(d) show four different regions of the embryonic roof and their insertion into a host embryo. The top embryo shows induction of sensory buds and mesenchyme; the second from top almost a complete head with brain, ganglia and a cyclopic eye; the third has ganglia and ear vesicles; and the fourth, a spinal cord and a limb.

Subsequent findings by other experimental embryologists soon called the organizer concept into question (Hamburger, 1988). A particularly annoying finding (for Spemann) was the discovery that boiled and hence dead organizer has a similar effect as live tissue, suggesting that the inducer was chemical in nature rather than a live process, as Spemann believed (perhaps due to Driesch's influence). Furthermore, various fractions of blended organizer tissue had the effect. While this is what one might expect if the inducing substances are chemical, there were also numerous unspecific or, as they were called, "heterologous" factors capable of mimicking the organizer phenomenon, for example, methylene blue or sand particles. This was more of a difficulty for those favoring an explanation in terms of specific chemical signals. There are also tissues from other embryonic parts or from other organisms that can have similar effects, even after boiling.

Today, it is known that Spemann's and Mangold's newt embryos are especially prone to showing such nonspecific or heterologous inductions, as are other urodeles (salamander-like amphibians with tails), in particular the Mexican axolotl. By contrast, tailless anurans such as the African clawed frog *Xenopus* show much fewer, if any nonspecific inductions. In any case, the whole organizer theory received a serious blow because the possibility of nonspecific induction seemed to imply that the organizing power was really in the responding tissue and that the organizer was a mere trigger. De Robertis (2006) even recalls that "[b]y the time I was a student in the 1970s, it was common to hear comments such as 'Spemann's organizer set developmental biology back by 50 years.'"

In a way, these difficulties were similar to the ones encountered with eye induction in the first decade of the twentieth century, mentioned in Section 2.1. It seems that, especially in urodeles, embryonic tissues can be coaxed by various more or less harsh treatments toward the initiation of major developmental pathways, even forming whole new body axes. This was a major challenge to the theory of embryonic inductions and led Spemann to propose the theory of "double assurance" (*doppelte Sicherung*). According to this idea, major developmental pathways such as the one leading to lens formation or neural development have backup mechanisms that can be released by nonspecific triggers in case the main control mechanism fails. Thus, perhaps what the embryologists were observing in the various graft experiments was merely the triggering of such backup mechanisms under conditions of stress. The theory of double assurance has recently been reinvigorated by molecular findings in *Xenopus*.

Spemann rather ingeniously defended the organizer concept against the challenge of nonspecific inducers. He first introduced the concept of an *action–reaction system* that includes the organizer or inducing tissue as well as the responding tissue. Then he suggested that developmental processes such as formation of the neural plate (which later develops into the neural tube) are normally the result of an interaction between inducing and responding tissue. In addition, some tissues have the power of triggering developmental events:

I always considered the inducing action of the organizer as a trigger. Furthermore, the question about the contributions of the action and reaction systems in the formation and in the nature of the product of induction has been discussed from the beginning. Experiments done in order to resolve this issue have attributed increasing importance to the reaction system; eventually it became so great as to call the very concept of organizer into question. (Spemann, 1936: 276, my translation)

As Spemann makes clear in this passage, he does acknowledge that the power of some responding tissues to self-organize while the organizer provides at most a trigger is a challenge to the very idea of an organizing center. Furthermore, he accepted that most of the "complication" (a term that can mean "complex mechanism" in German; for example, in Swiss watchmaking) may reside in the responding tissue. At the same time, he insists:

However, this does not annihilate the experimental result that a piece of upper blastopore lip invaginates in the direction that corresponds to its inner structure [*seinem inneren Bau entsprechend*], even against the host's axes; that it supplements [*ergänzt*] itself from its mesoderm environment to a complete system of axes; that a piece of archenteron [primary gut, M.W.] roof inserted into the blastocoel and hence below the ectoderm induces a medullary [neural, M.W.] plate, which can be perpendicular or in opposite direction to the primary plate, and that it arranges lenses and ear vesicles in the right order and proportion. (276)

Spemann concludes:

Apparently, induction by an inducer without morphological structure differs in one important point from the one caused by a living organizer; to wit, that direction and kind of change are determined alone by the reaction system, in its inner structure and its other states. (276)

Thus, Spemann's main point is that in the case of a live organizer graft, it is the graft that determines the orientation and position of the induced secondary body axes, while in the case of heterologous inducers it is the host system itself that determines these parameters.

What is the significance of Spemann's point? I will argue in Section 2.3 that this defense of the organizer concept by Spemann contains an important lesson about causality in developmental biology.

2.3 Organizer Grafts as Experimental Tools for Causally Specific Interventions

What exactly did classical experimental embryologists mean by "nonspecific inductions"? The most straightforward reading of this idea is in the sense of what Woodward (2010) has termed "one-to-one specificity." This kind of specificity designates causal relations where there is a range of different causes and a range of different effects and, within these ranges, one cause has only one effect and any effect has only one cause (see Section 1.5). (In this context, "cause" does not mean sufficient cause but "causally relevant factor in the circumstances.") A weaker version of this kind of specificity is present when there is one cause from a range of causes that shows a preference for one or a few of its effects. In biochemistry, an example of such a preference is selective binding affinity of a molecule (e.g., an enzyme) for one or a few of its possible ligands or substrates (Lean, 2019). This property, also known as binding specificity, is a prerequisite for life as we know it. Some of the heterologous inducers discovered by classical experimental embryologists (see Section 2.2) were clearly nonspecific according to this sense of specificity (i.e., they can have many different effects other than inducing lenses or neural tissue).

Why were unspecific inducers considered a problem? Would classical embryologists have been happier had they found a substance or a mixture that was specific for its effect or for a binding partner in the one-to-one sense? The answer is clearly yes. As Lean (2019) reminds us, the precise regulation and organization of biological processes can only be explained by interactions that are specific in the one-to-one sense. Furthermore, specific interactions are of enormous heuristic value for experimental interventions because they allow experimentalists to manipulate just one or a few variables in a system without affecting others ("surgical" versus "fat-handed" interventions). I will come back to the significance of this sense of specificity in Section 2.4.

In any case, there is another sense of "specificity" that is relevant here, a sense that is useful for bringing out the scientific significance of Mangold and Spemann's discovery. According to Woodward (2010) and Waters (2007), a causal link between a variable C and a variable E is *causally specific* to the extent to which there are lots of different values of C and of E and the values of C and of E map onto each other in a one-to-one fashion (i.e., bijectively or nearly so). This is supposed to capture the idea of a cause enabling fine-grained control over its effect. In this sense, a light dimmer is more causally specific than a mere on–off switch. The most discussed example of causal specificity in the philosophy of biology literature is nucleic acids (DNA and mRNA) and their effect on protein sequence (Griffiths et al., 2015; Waters, 2007; Weber, 2006, forthcoming-a). Clearly, this sense of causal specificity is distinct from one-to-one specificity, but it is also a potentially relevant feature that causal relations may have. I will show now that organizer grafts are distinguished by enabling causally specific control in the fine-grained sense over developmental processes. Thus, causal specificity in the fine-grained sense is not a unique property of linear biomolecules (DNA, RNA or protein), as the previous debate over causal specificity (Griffiths et al., 2015; Waters, 2007; Weber, 2006, forthcoming-a; Woodward, 2010) might suggest. We can find it also at the tissue level.

The idea is that organizers (or organizer grafts) can be represented by a causal variable that affords causally specific control in the fine-grained sense. As Spemann explained in his defense of the organizer concept (see Section 2.2), an organizer graft determines the orientation of the secondary axis it induces, including the direction in which the neural plate forms and the place and proportions in which lenses and ear vesicles form. Thus, an organizer graft allows an experimenter to spatially control some developmental processes in a fine-grained way, meaning that there are several different ways of inserting a graft that cause different outcomes of the developmental processes. By contrast, as Spemann insisted, the heterologous inducers afford no such control. In such cases, the spatial organization is controlled by the host, while the heterologous inducers just act as triggers or switches.

Spemann was also able to show that the organizer has *temporal* specificity. When he transplanted organizer tissue from an early gastrula, it induced the formation of a secondary head, while tissue taken from a late gastrula would form a tail. This gave rise to the notions of a "head" and a "trunk" organizer (Hamburger, 1988: 61). Spemann's colleague Otto Mangold (Hilde Mangold's widower) was even able to demonstrate *regional* specificity of organizer grafts, as different regions from the archenteron roof (invaginated blastopore lip that moved to the opposite side inside the embryo) induced different structures upon insertion into the blastocoels of newt gastrulae. He sectioned the archenteron roof (i.e., organizer tissue after invagination) into four different regions or segments and showed that each one of these segments, upon insertion into salamander gastrulae, caused different mesoderm and neural structures to emerge in the host embryo (see Figure 1). This regional specificity, too, can be understood as causal specificity in the fine-grained sense, because different values of the cause variables (here, different regional parts of the organizer) cause different values of the effect variable.

I suggest that construing organizers and in particular organizer grafts as well as other inducers as sites of causally specific interventions in the fine-grained sense allows us to fully grasp their scientific significance as well as their heuristic value for subsequent research. This value will be elaborated in Section 2.4.

2.4 On the Discovery Value of Specific Causes

We have seen in Section 2.3 that experimental embryologists of the pre-molecular era had identified sites for experimental intervention that enabled them to control developmental processes in a fine-grained way. Organizer grafts and other tissues with inducing powers differed from other ways of releasing developmental pathways (e.g., heterologous inducers) in their causal specificity with respect to different possible outcomes of developmental processes (e.g., the position and orientation of secondary body axes or the embryonic structures induced by different regions or time slices of an organizing center). Causally specific control in the fine-grained sense seems to distinguish certain grafts with inducing powers such as the Spemann–Mangold organizer from heterologous inducers, which only seem to be able to switch on a process that is controlled by other factors. Organizer tissue can act as a switch, but it can also exert causally specific control in the fine-grained sense. This double role that organizer tissue can play has been extremely confusing, but there is no conceptual difficulty here.

Philosophers of science have argued that causal specificity in the fine-grained sense is a causal selection criterion (see Section 1.5). Indeed, it can be argued that the organization and coordination of complex biological processes require at least some causal variables with fine-grained control. Thus, knowledge of specific causes in this sense has *explanatory* value. However, there may be additional reasons for focusing on such causes. I would like to suggest that causally specific causes in the fine-grained sense can also have a considerable *heuristic* value for scientific discovery.

To appreciate the discovery value of causal specificity, let us briefly consider how the first molecules implicated in the organizer phenomenon were identified several decades later. In 1992, the laboratory of Edward De Robertis identified a gene in the frog *Xenopus* the messenger-RNA product of which mimicked some of the effects of the Spemann–Mangold organizer upon injection into frog embryos. (Messenger-RNA or mRNA is an intermediary in the synthesis of proteins from genes.) They named the gene *goosecoid* because it showed a sequence resemblance (a "homeobox"; see Box 3) to two known genes from *Drosophila*, *gooseberry* and *bicoid*. The way in which this gene was identified is remarkable, because the researchers started with mRNA isolated from the blastopore lip of *Xenopus* embryos, which is where the organizer is located. Of course, they chose this region because they knew about its significance from Spemann and Mangold. Newly available recombinant DNA technology allowed them to obtain DNA sequences complementary to these mRNAs. Finally, the homeobox sequence's similarity to known *Drosophila* genes allowed them to isolate the *goosecoid* gene. It turned out to be active precisely in the organizer region (Cho et al., 1991).

A similar approach was taken to isolate several other genes the protein products of which seem to mediate the effects of the organizer (e.g., a protein named Chordin). This is a protein that blocks the activity of certain growth factors that determine the fate of some cells. A crucial finding was the existence of a cocktail of proteins secreted by the organizer that establishes a gradient of growth-factor signaling activity along the dorsoventral (DV) axis of the *Xenopus* embryo (similar gradients were found in other animals as well such as the zebrafish). This signaling gradient patterns the embryo along the DV axis, determining whether the cells later become epidermis, neural tissue, noto-chords, skeletal muscle, kidneys or blood vessels (see Box 1).

In facilitating the discovery of the genes and gene products such as the proteins shown in Box 1 and their interactions, the causal knowledge as well as the techniques inherited from classical experimental embryology played a crucial role in identifying the molecular growth signals produced by the organizer as well as other induction mechanisms (see Weber, forthcoming-b for a more detailed account).

What makes knowledge about specific causes in the fine-grained sense such as organizer tissue useful for learning about molecules involved in the control of development? Let us begin to address this question by first considering the heuristic value of one-to-one specificity. As Lean (2019) has argued, binding specificity can be an experimentally useful feature for making sure that inter-ventions in a biological system change only one (or not too many) variables at once. There are many cases where binding specificity is a feature that scientists can exploit to learn about biological processes. Another example is the use of gene sequences isolated from *Drosophila* in order to identify developmental genes in other species, which was made possible by the fact that some develop-mental genes share highly conserved sequences such as the "homeobox" (Gehring, 1998). The DNA probes used to fish for related genes in tissue samples from other species such as *Xenopus*, mice or humans show an extremely high binding specificity to some DNA sequences in the target species, which was instrumental for their isolation.

I don't want to suggest that causal specificity in the fine-grained sense is as useful a property of causal relations as binding specificity. After all, any interven-tion that has a graded response, such as a dose–response relationship,[8] is more or

---

[8] It should not be forgotten that the presence of a clear dose–response relationship or a "biological gradient" is one of the famous Bradford Hill criteria that are considered to be indicative of causal relationships in epidemiology. However, the criterion alone is clearly not sufficient for causality. In any case, we are here not so much interested in the question of how causal relationships can be proven but rather how certain features of established causal relationships, in particular specificity in the two senses under consideration, can facilitate scientific discovery.

Box 1 The Chordin/Bone Morphogenetic Protein Signaling Pathway

(Reprinted from De Robertis and Moriyama (2016) with permission.)

Some of the effects of the Spemann–Mangold organizer are today attributed to a cocktail of proteins which are secreted by the organizer cells and which diffuse away from these cells to control the growth and differentiation of other cells located at some distance from the organizer. Some of these proteins, including a group of growth factors named bone morphogenetic proteins (BMP2/4/7 and ADMP in diagram C shown in the figure), are capable of binding to specific receptors on the cell surface, which then send a biochemical signal to the cell nucleus. These signals are transduced in the form of cascades of phosphorylation reactions that eventually lead to the activation or inactivation of transcription factors in the cell nucleus (see also Box 2). Transcription factors are DNA-binding proteins that specifically regulate the expression of various genes. Chordin protein, secreted at a high concentration by the Spemann–Mangold organizer, binds and inactivates BMPs. Tolloid is a proteinase, an enzyme that specifically breaks down Chordin. Sizzled inhibits Tolloid. Furthermore, Chordin was shown to shuttle along the DV axis (shown as "flux" in diagram C; see also Section 3.4). Thus, a dynamic process of protein secretion, transport, degradation and feedback creates a gradient of BMP-signaling activity along the DV-axis, with the signaling activity being highest at the ventral side, lowest at the dorsal side (where the organizer secretes the BMP-antagonist Chordin) and intermediate in between. This signaling gradient determines different cell fates along the CV axis and can be visualized at different developmental stages. In photographs A and B in the included figure the gradient has been visualized by the activity of an intermediate in the signaling pathway, a phosphatase called Smad (see Box 2 for the role of Smad). The brightly lit areas show that BMP signaling activity is highest in the regions far from the organizer and therefore from the BMP-antagonists it secretes.

less specific in the fine-grained sense. In spite of this caveat, I suggest that there are circumstances in which the identification of a causal variable enabling a certain kind of fine-grained control over processes is an important starting point for learning more about the regulation of these processes. In developmental biology, where the spatial arrangement and timing of events are crucial, causal variables such as organizers and other tissues with inducing powers that enable fine-grained control over spatial arrangements and timing are key to finding the regulatory mechanisms that control these processes, such as the one shown in Box 1. And this turned out to be the case, as witnessed in particular by the Spemann–Mangold organizer, which led scientists to the signals that govern embryonic cell differentiation (see also Section 3.4).

2.5 The Organizer Goes Molecular: Reduction or Replacement?

Is there a sense in which the body of knowledge known as classical experimental embryology, some central parts of which I have reconstructed in this section, has been reduced to molecular biology? The identification of numerous genes and gene products (mRNA, proteins) that seem to be responsible for at least some of the phenomena studied by classical experimental embryologists such as Hilde Mangold, Hans Spemann and others (see Box 1) seems to suggest that much.

A crucial question is whether the classical body of knowledge has been *reduced* or rather *replaced*. This question has been widely discussed in the context of other examples, including the case of classical thermodynamics/ statistical mechanics and classical genetics/molecular biology (Feyerabend, 1962; Schaffner, 1993). Most of the interesting cases seem to lie somewhere in between a complete reduction in the sense of entailment (see Section 1.2) and a complete replacement of one theory by an incompatible one. Thus, if there is a reduction, the reducing theory usually corrects some errors in the theory to be reduced (e.g., classical thermodynamics said that thermic change with negative entropy is impossible while statistical mechanics allows it in principle, albeit in bulk matter and only with extremely low probability).

The developmental biologist Scott Gilbert (2001) has provided a very illuminating discussion that is relevant to our question. He points out that, by 1980, most developmental biologist had given up believing that the molecules involved in embryonic induction would ever be identified. As we have seen in Section 2.4, this changed dramatically with the arrival of recombinant DNA technology that was able to exploit sequence similarities between *Xenopus* and *Drosophila* in order to identify genes the protein products of which are involved in phenomena such as the Spemann–Mangold organizer; see also Weber (forthcoming-b). However, what molecular developmental biologists found with their new, powerful tools wasn't always quite what they expected.

The dominant model of the action of the Spemann–Mangold organizer had always been that the mesoderm cells making up the organizer do something to the ectoderm cells that lie above them, probably by secreting a chemical signal so as to change their fate from epidermis to neural tissue. A surprise discovery stood this idea on its head: Ali Hemmati-Brivanlou and Douglas Melton (1997) found that the inactivation of a receptor for certain growth factors by genetic manipulation of the corresponding receptor genes in *Xenopus* led to the appearance of another protein molecule that is expressed only by neural cells. This suggested that it was the *in*activation of certain growth factors by proteins secreted by the organizer (see Box 1) that caused cells to turn neural. Further to this finding, they were able to demonstrate that, prior to receiving any molecular signals from the organizer, cells were already committed to become neural cells; however, they were blocked from following this path by some proteins called bone morphogenetic proteins or BMPs (see Box 1). Then, proteins secreted by the organizer antagonize the BMPs so as to allow the cells to realize their neural fate. Thus, it looks like it was exactly the other way around from what Spemann and colleagues had always thought: epidermis is induced, while neural is the default pathway.

It should be noted that the molecular account differs more radically from the classical one than it might seem, because it changed not only the attribution but also the definition of the default state. When classical embryologists thought that epidermis is the default and neural the induced state, they meant this: *default = cell fate in the absence of organizer (or other inducing) activity.* What molecular biologists mean when they say that BMPs block the default neural pathway and that organizer proteins remove that block is: *default = cell fate in the absence of specific molecular signals present before organizer /*

inducer activity sets in.[9] Thus, the default state is not defined in the same way by classical embryologists and molecular developmental biologists. What is more, the operational criteria for identifying the underlying dispositions also appear to be different: in classical embryology, the standard experimental protocols for determining cell fates involved transplantation and (somatic) genetic mapping; in molecular biology, the detection of molecular markers for certain differentiated cell types (e.g., neural markers).

Thus, it is not the case that scientists simply described a causal role at the higher level and then discovered its molecular realizers, which is how some philosophers have tried to describe explanatory reduction (Kim, 2007). The molecular findings changed the causal role attributed to the organizer. This could mean that we have incommensurable concepts in the sense of Kuhn (1970) here, which could be an indication that the molecular explanation *replaces* rather than *reduces* the classical account.[10] Such an interpretation would also be well in line with Gilbert's (2001) description of the case as a "paradigm shift," a notion famously introduced by Kuhn.

In spite of the paradigm shifts invoked by Gilbert, there is also much continuity across the molecularization of developmental biology (which is fully compatible with Kuhn's account). Gilbert sees it in the large number of citations to the works of classical embryologists found in publications of molecular work. I contend that the continuity mainly lies in specific experimental manipulations developed in the classical era that proved to be invaluable for identifying the molecular signals responsible for embryonic inductions; see Section 2.4 and, in more detail, Weber (forthcoming-b). In some ways, the case also resembles that of classical genetics, where molecular biology has led

---

[9] Spemann also thought that the organizer was a "primary" inducer, meaning that it had its own fate autonomously determined while all other tissues had to be induced. This is no longer considered to be true; the organizer is itself induced by the so-called Nieuwkoop Center. This was established with classical methods, so it wasn't part of the molecular paradigm shift.

[10] The Kuhnian concept requires that the incompatibility of the incommensurable theories not be reducible entirely to logical contradictions (Hoyningen-Huene, 1990). Indeed, we could say in our case that Spemann's account of neural induction wasn't simply false from the contemporary point of view (Edward De Robertis told me in a personal conversation that Spemann was "pretty close"). The organizer *does* cause neuralization of ectoderm; it's just that it doesn't do so by initiating the neural differentiation process but by removing a block to a process that had already been initiated. There might even be some resemblance here to the well-known case of phlogiston chemistry (one of Kuhn's own examples of incommensurability): Air said to be "enriched in phlogiston" does suffocate animals, but the real cause is not that the air had been enriched in phlogiston but that it had been depleted of oxygen. Another possible parallel to a Kuhnian example is Aristotelian physics, which has a different conception of force and motion according to which there can be no motion without force. In Newtonian physics, such motion is possible in principle (inertial motion). In all these cases, the phenomena are classified differently by the incommensurable accounts.

to what Waters (2008) termed a "re-tooling" of the investigative practices of the classical discipline rather than the reduction of a core theory.

3 Idealized Causal Concepts as Research Tools: Morphogens

As we have seen in Section 2.5, some of the molecules that are involved in embryonic induction form gradients along an embryonic axis, and the induced tissues may respond differently to different concentrations or activity levels of these substances. Such substances are also known as "morphogens" because they control the generation of embryonic patterns.[11] In Section 3.1, I will give a brief overview of the development of morphogen theory and its experimental confirmation. Then, in Section 3.2 I discuss what I consider to be an idealized causal model of morphogen action, namely the so-called French Flag Model. In Section 3.3, I explore the heuristic role of this model and the underlying morphogen concept as a tool for scientific discovery. Finally, in Section 3.4 I examine to what extent the discovery of morphogens has solved the puzzles about morphogenetic fields raised by Driesch, Spemann and other classical embryologists, and how my causal perspective can illuminate this discovery.

3.1 From Theory to Experimental Confirmation

According to an idea that goes back to the beginning of the twentieth century, early embryos contain either concentration gradients of some substance or gradients of metabolic activity that determine a cell's later fate as a function of its location within the embryo (Rogers and Schier, 2011). Early evidence for such morphogen gradients came in particular from studies of the remarkable regenerating power of planaria (flatworms). Eerily, even tiny fragments of a flatworm can regenerate the whole worm. Interestingly, the fragments retain their initial polarity (i.e., the part of the fragment that was closer to the head will form a new head with a rate that depends on the distance from the original head). A possible explanation for this would be a concentration gradient of some substance or of some metabolic activity that determines the future development of different parts of the worm. Such ideas were proposed at the beginning of the twentieth century by Thomas Hunt Morgan (later of *Drosophila* genetics fame) and the zoologist Charles Manning Child.

In the 1950s, the highly influential mathematician Alan Turing constructed a model in which two "morphogens" (he introduced this term) are synthesized

---

[11] Morphogens should be conceptually separated from what is called "morphogenesis" in developmental biology, which refers to the actual mechanical and hydraulic forces that push, pull and squeeze the embryo into shape. The question of how the mechanisms of morphogenesis and those of the genetic control of development (examined here) are interrelated is not trivial but cannot be addressed here. (See Love [2020a] for an illuminating discussion.)

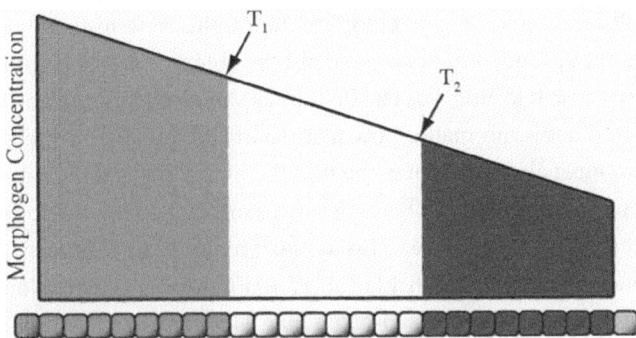

Figure 2 The French Flag Model as rendered in Jaeger and Martinez-Arias (2009). Image licensed under the Creative Commons Attribution.

and degraded at different rates, and one of the morphogens is converted into the other at a specific rate. Turing showed that such a reaction-diffusion system would be able to generate a wavelike pattern along a single dimension, thus providing a potential explanation for the origin of embryonic patterns.

Another influential idea is the theory of "positional information" proposed by the embryologist Lewis Wolpert (with acknowledgements to Driesch's ideas). Wolpert wanted to formulate a general account of "how genetic information can be translated in a reliable manner to give specific and different spatial patterns of cellular differentiation" (Wolpert, 1969: 1). He assumed that, just as there is a general account in molecular biology of how genetic information directs "molecular differentiation" (i.e., the synthesis of proteins), there would be a "universal mechanism" by which genetic information is translated into spatial patterns of differentiation. But this kind of genetic information would reside not in the nucleotide sequence of DNA but in the concentration or activity of some substance, not identified by Wolpert (see Figure 2). His basic idea was that there would be some mechanism whereby the differentiation state of a cell responds to the cell's position with respect to a set of points in the system. Thus, Wolpert pretty much postulated an abstract entity – positional information – to do the job of Driesch's entelechy (Section 1.1).[12]

Wolpert also devised the well-known "French Flag Problem." The problem is the following. Wolpert imagined a developing system that must form a pattern resembling the French flag. Now, the problem is not only that something must

---

[12] Indeed, a charitable interpretation of Driesch consists in crediting him with offering the first theory of positional information. (Note the similarity of Wolpert's theory of positional information to Driesch's formal account of entelechy in Section 1.1.) While one could view information as something immaterial, thus vindicating Driesch, this would still leave his problematic arguments against any mechanistic explanation (see Section 1.1), which were not endorsed by Wolpert.

"tell" the different parts what "color" to adopt, this system must generate the same pattern independently of size and also be able to readjust if some parts are removed (because real embryos can do this, as shown by Driesch and others; see Section 1.1). That is, no matter how many parts are removed from the system, the left third must always be blue, the middle third white and the right third red. Furthermore, the system must be scale-invariant, that is, it must form the same pattern irrespective of its size. This is the French Flag *Problem*. It must be distinguished from the French Flag *Model* (Figure 2), which is an idealized model of morphogen action. The French Flag Model by itself is not a *solution* to the French Flag Problem; it merely provides a way to *state* the problem.[13] For to explain the robustness of pattern formation it must be shown that the mechanisms that generate and interpret the gradient are able to re-equilibrate in a way that preserves the pattern after a disturbance. Wolpert put the problem like this: "Pattern regulation, which is the ability of the system to form the pattern even when parts are removed, or added, and to show size invariance as in the French flag problem, is largely dependent on the ability of the cells to change their positional information and interpret this change" (Wolpert, 1969: 1). Thus, the gradient must be able not only to buffer disturbances but also to adapt to the size of the system. This would come for free if the gradient were somehow able to adapt to the tissue size. As we shall see in Section 3.4, this problem requires computational modeling.

Remarkably, most of the theoretical ideas developed in the twentieth century about morphogens and positional information are still in use in some form or another in contemporary developmental biology. However, clear experimental evidence for the operation of such mechanisms of pattern formation as they have been postulated had to await the 1980s. Three model organisms (*Drosophila*, *Xenopus* and zebrafish) as well as recombinant DNA technology proved to be game changers in this quest.

The first morphogen to be experimentally confirmed was the protein Bicoid, found to form a concentration gradient in *Drosophila* (Driever and Nüsslein-Volhard, 1988). During egg formation, *Drosophila* mothers deposit a maternal mRNA at the front end of an oocyte. After fertilization, this mRNA is used to synthesize the Bicoid protein that subsequently forms an anteroposterior (front-to-end) gradient. By manipulating the Bicoid concentration gradient with

---

[13] In fact, Wolpert never used the term "French Flag *Model*" and discussed different possible solutions to the French Flag *Problem* in an earlier article, including solutions that do not require a gradient. It is thus important to distinguish between Wolpert's general *theory* of positional information (his postulate of positional information and its role in development), the French Flag *Problem* which arises within that theory and specific *models* of how positional information is realized in an embryo.

genetic methods that changed the copy number of the corresponding gene, the fly biologists obtained evidence that this gradient determines the anteroposterior location of certain embryonic structures as well as gene expression patterns later on in development, including the boundaries between the different segments that mark the insect body or the boundary between the head and thorax (see Figure 3). Thus, like in the case of regional induction specificity by

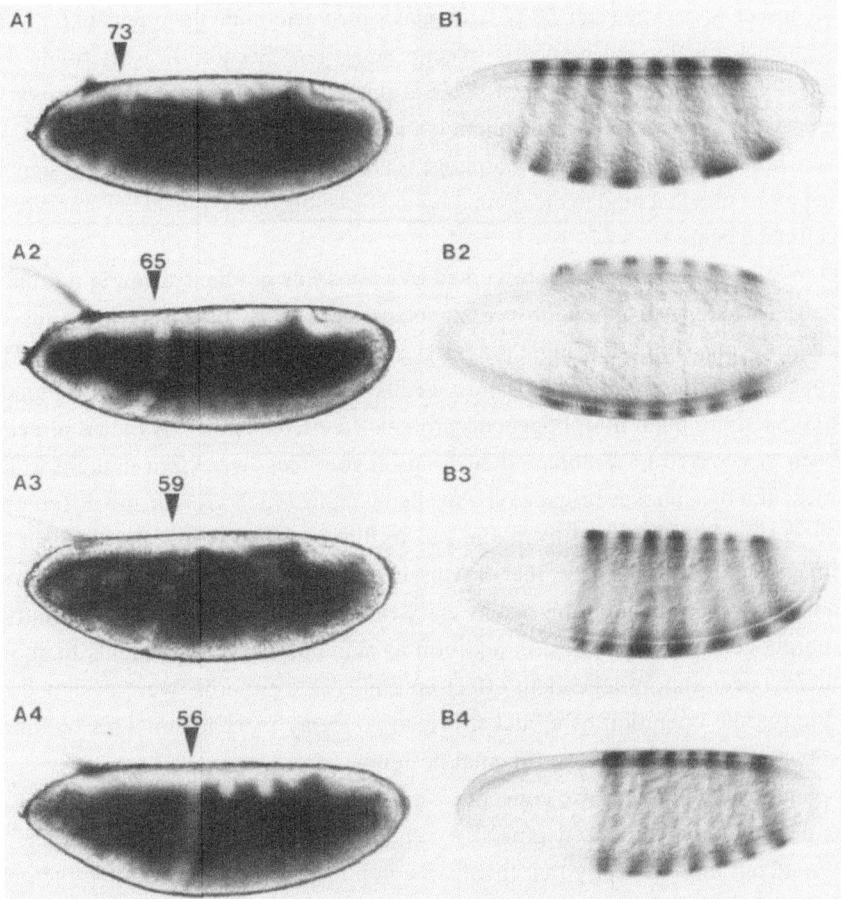

Figure 3 Effect of genetically manipulating the Bicoid morphogen gradient in *Drosophila* embryos. The arrows on the left show the location of the head fold, which is shifted in the posterior direction as gene dosage increases. The stripes on the right show the expression pattern of another gene, *even skipped*, which also shows a posterior shift in response to altering *bicoid* gene dosage (and hence Bicoid protein concentration). Image reproduced with permission from Driever and Nüsslein-Volhard (1988).

organizer implants (Figure 1), these experimental interventions exhibit causal specificity in the fine-grained sense, the significance of which I have explained in Section 2.3. We can also see the manipulationist criterion for causality (Section 1.5) in action here.

Bicoid turned out to be a transcription factor that directly binds to DNA and controls the activity of numerous other genes along the anteroposterior axis (the "gap genes"). Interactions between these genes lead to the periodic expression of yet another set of genes (the "pair-rule genes") that define the segment pattern of the insect larva. The segment polarity genes then determine the orientation of the segments. Finally, the homeotic selector genes (see Section 4.3) determine the identity of each of the segment. This is possible only because the *Drosophila* embryo at this stage is a single but polynucleated cell with thousands of nuclei. In vertebrates, gradient-forming morphogens are not transcription factors but signaling molecules that bind to membrane-bound receptors on the outside of embryonic cells (see below).

While Bicoid was rapidly accepted as a bona fide gradient-forming morphogen, it was initially thought to be a special case that works only in an insect embryo consisting of a single polynucleated cell, but not in an embryo already consisting of hundreds or thousands of cells. While some morphogenetic signals (e.g., activin, bone morphogenetic proteins (see Box 1) or WNT) had already been discovered in *Xenopus*, their long-range effects were first thought to be mediated by a bucket-brigade type mechanism, in which growth and differentiation signals are relayed from one cell to the next. It was initially difficult to imagine how a substance that travels in the extracellular space could form a gradient by diffusion. This is why early experiments trying to confirm putative morphogens in vertebrates, in addition to manipulating the gradients to show their concentration-dependent effect on embryonic patterns, were designed to rule out the operation of a bucket-brigade or relay mechanism. This required some rather ingenious experimental designs.

Thus, it seems that the concept of gradient-forming morphogens has been experimentally confirmed almost a century after its inception. However, it turned out more recently that there are complications with this simple picture. I will show now that we can best make sense of the French Flag Model by attending to its idealized nature.

3.2 The French Flag Model of Morphogen Action as an Idealized Causal Model

Let us consider what the simple French Flag Model, as drawn in Figure 2, posits in terms of causal relations. We can treat the morphogen concentration as a continuous causal variable [*M*], which is a function of time *t* and relative position *r* in the embryo: [*M*] = *f*(*r*,*t*). This variable takes different values along the embryo's anteroposterior axis and also varies in time. (It starts as a very steep gradient, then flattens by diffusion and possibly other mechanisms and eventually disappears.) What [*M*] does according to the model is to cause differences in cell fate along the anteroposterior (AP) axis. Initially, these cells have the same fate. But after the morphogen has acted, their fate differs.

Technically, according to the model, the morphogen is *the actual-difference making cause* (Waters, 2007) of cell fate in the population of cells (or cell nuclei in the case of Bicoid) that are located along the anteroposterior axis. What this means is that the actual variation in [*M*] alone causally accounts for the differences in cell fate in the population of cells located along the AP axis. This doesn't imply that [*M*] is the *only* factor that is causally relevant to cell fate or that it is by itself causally sufficient for cell fate. Quite on the contrary, what the model supposes is that there exist additional causal factors that "interpret" the gradient, which must also eventually have an effect on cell fate. But those other factors *do not vary* along the AP axis, at least not initially. In this sense, it is variation in [*M*] that accounts for the later differences in cell fate. In fact, it *fully* accounts for these differences, because if the gradient were abolished in a way that changed no other variables relevant to cell fate, all differences would disappear, at least according to the model. (In real morphogen systems this doesn't always appear to be the case.) In the simple French Flag Model, there is no other causal variable that actually varies and that could be manipulated in such a way as to make all differences in cell fate disappear. Thus, Waters's (2007) conditions for the actual-difference making cause are satisfied; but (unlike in Waters's own examples) it will turn out to be an idealization, as I will show in this section.

I suggest that this analysis captures the causal content of the French Flag Model and provides an explication of the notion that morphogens "pattern fields of cells" (Ashe and Briscoe, 2006), even though they are far from being causally sufficient for any pattern. I would like to go even as far as to suggest that this causal analysis also provides an analysis of the very *concept* of a morphogen, which can only be understood as an idealization.

Another aspect of the causal role of morphogens is their *causal specificity* in the fine-grained sense that we already encountered in Section 2.3 (i.e., in the sense that there are many states of the cause-and-effect variables and a causal dependence that allows fine-grained control of the latter by the former). Are morphogens causally specific in this sense? Well, what seems to be clear is that morphogens as construed in Wolpert's model are not simple two-state switches; rather morphogens are thought to effectuate a choice between at least three different states. However, one of Woodward's (2010) conditions for causal specificity does not appear to be satisfied, namely the condition that there be not too many distinct states of the cause variable that map onto the same state of the effect variable (i.e., which is part of the condition of near-bijectivity of the mapping). Rather, according to the simple model, there are many states of the morphogen concentration variable that give the same cell fate. However, there is, according to the French Flag Model, a bijective mapping of three concentration *ranges* into three cell fates. Morphogens are not just switches. Furthermore, the model implies that there are interventions on the gradient that move compartment boundaries along the embryonic axis defined by the morphogen. Such positional shifts have actually been observed (see Figure 3). But as morphogen concentration is a continuous variable whereas cell fate is not, a causally specific mapping in Woodward's and Waters's sense also requires a coarse-grained description of morphogen concentration.

This still leaves the question of how we should understand the talk about positional information. The use of information concepts in biology has been the subject of much debate in the philosophy of biology (Godfrey-Smith and Sterelny, 2016), and I cannot engage with this complex debate here. One of the central issues in this debate is the question whether information should be construed in terms of semantic content of the kind that thoughts or language have. I don't think this has to be assumed here, as there is a straightforward way of understanding the notion of positional information, namely in terms of a correlation between the morphogen concentration and relative position. Thus, the mere claim that morphogens encode positional information is just an abstract way of saying that there are mechanisms that establish a sufficiently reliable correlation between relative position and morphogen expression level.[14] The strength of this correlation can be measured by using mathematical information theory (Dubuis et al., 2013).[15] Of course, as correlations are symmetrical, this construal doesn't account for the directionality of concentration-dependent cell fate determination; a causal condition has to be added for this purpose.

Now, I think it was always clear that this model, as depicted in Figure 2, isn't very realistic; it contains numerous *idealizations* and *abstractions*. Following standard terminology, I shall mean by "abstraction" the omission of details from the model, and by "idealization" the deliberate supposition of falsehoods in the

---

[14] Levy (2011) gives a similar interpretation of the notion of positional information, except that he considers information talk to be fictional. In my Weber (2005: Chapter 8), I also took the view that positional information should be understood *in analogy* to semantic information, but without semantic content.

[15] Love (2020b) argues that developmental biologists use information theory to measure the causal specificity of the morphogens with respect to cell fate in order to determine if the morphogens can account for the phenomena.

interest of tractability or computability (such as frictionless motion or point particles in mechanics).

The French Flag Model *abstracts* from:

(1) processes that link the morphogen gradient to gene expression patterns in the recipient cells; it merely supposes that such processes exist.

(2) the temporal dimension in looking only at the causal dependencies of cell fate on the morphogen concentration. (Perhaps it doesn't abstract completely from temporality in supposing that the gradient is formed *before* the cells become committed, but it disregards any events that happen in between.)

While (1) seems to be a clear case of an abstraction, (2) looks more like an idealization, because morphogen systems could be dynamic, in which case time cannot be omitted on pains of distorting reality (see Section 3.3). Abstracting from time works only if the formation and the interpretation of the gradient are separate processes that don't interact (perhaps because the interpretation only begins once the gradient has formed). But this could be false, as we shall see (Section 3.3).

In addition, there are the following assumptions, which also have the ring of idealizations:

(3) Only three qualitatively distinct cell fates.

(4) Determinism (i.e., all cells react to a given morphogen concentration in the same way).

(5) Sharp morphogen thresholds (as opposed to downstream causal interactions) account for the precision of compartment boundaries.

(6) No variation in the morphogen gradient between individual embryos; developmental noise is neglected.

(7) No other factors that actually vary affect cell fates (i.e., the morphogen is *the* actual-difference making cause of cell fate in the sense of Waters (2007) and *fully* accounts for differences in these).

(8) The generation and interpretation of positional information are independent processes.

(9) The morphogen acts only via its concentration and not via additional properties (e.g., exposure time).

(10) Different body axes are patterned independently.

Most of these assumptions, while not impossible (unlike point particles), were never very likely, given what embryologists already knew when Wolpert proposed the model; so it is clear that the model was never intended to be a faithful representation of the world. It's an *idealized causal model*. I will discuss in Section 3.3 what idealizations the model incorporates and how these idealizations guided research.

3.3 The Morphogen Concept as a Research Tool

From the foregoing discussion it is obvious that there are various ways in which the simple model depicted in Figure 2 could be enriched in detail as well as de-idealized.

(1) Of course, biologists are interested in identifying the targets of the morphogens, or, as they often express themselves, to determine "how a graded signal is transformed into alterations in gene expression programs, such that the positional information supplied by the morphogen produces the appropriate spatial pattern of cellular differentiation" (Ashe and Briscoe, 2006).

(2) Time could be introduced by developing dynamical models.

(3) The model can be expanded to more than two thresholds; actually, up to seven have been identified in the dorsoventral axis of *Drosophila*.

(4) Instead of deterministic causality, the model could be turned probabilistic, such that variation in [M] would not account for all the variation in cell fates but only for their probability distribution. In fact, this fits nicely with information-theoretic treatments of positional information (see Section 3.2).

(5) The assumption that precise thresholds account for precise boundaries could be relaxed by elaborating "correction" mechanisms that account for precision.

(6) Interindividual variations in gradients due to noise could be taken into account.

(7) Finally, the morphogen could be construed as accounting not *fully* but only *partially* for differences in cell fate along the embryonic axis in question. Technically, this would correspond to a replacement of "the" actual-difference making cause by "an" actual-difference making cause in the sense of Waters (2007).

(8) Interactions between the generation and interpretation of positional information could be taken into account.

(9) The morphogen could be attributed with additional causally relevant properties (e.g., exposure time).

(10) Interactions between different axial patterning systems could be taken into account.

It turns out that several of these de-idealizations have actually been introduced in research on morphogens.[16] Let us begin with idealization (7), the uniqueness of the morphogen as actual difference maker.

---

[16] De-idealization is not merely a reversal of an idealization, as Knuuttila and Morgan (2019) argue; it involves the creation of new models.

For the case of Bicoid, it turned out that there are several repressors of Bicoid action that form gradients, and the dynamic interactions of these with Bicoid are necessary for the precise determination of compartmental boundaries in the insect embryo (Roth and Lynch, 2012). Such dynamic interactions also explain the scale invariance and robustness of the gradients (see also Section 3.4). It is not the case that spatial variation in Bicoid alone *fully* accounts for differences in cell fate; assumption (7) is false. In technical terms, none of these morphogens is thus *the* actual-difference making case of cell fate; each one of them is only *an* actual-difference making cause in the sense of Waters (2007). These concepts differ in that "the" actual-difference making cause fully accounts for some actual differences in a population (in the sense explained in Section 2.2), whereas "an" actual-difference making cause only partially accounts for it.

However, there may also exist morphogens that satisfy condition (7), for example, Chordin in *Xenopus* (De Robertis and Moriyama, 2016; see also Sections 2.2 and 3.4). We might therefore distinguish between *full* morphogens and *partial* morphogens depending on whether they fully or partially account for variations in cell fate. It might also be possible that some morphogen gradient once was a full morphogen in some lineage but due to evolution changed to a partial morphogen.

Another assumption that was shown to be false at least in the case of Bicoid is (5), that sharp thresholds account for the precision of compartment boundaries. Jaeger et al. (2004) used a dynamical modeling approach called the "gene circuit method" in order to determine whether the morphogen gradients of Bicoid, Caudal and Hunchback were able to account for the precise expression domains of some of their target genes. It turned out that they are not; in fact, some domain boundaries were shown to shift as the interactions between morphogens and targets unfold. Jaeger et al. (2004) concluded that, therefore, Bicoid, Caudal and Hunchback "do not qualify as morphogens in the strict sense." I think we can understand this "strict sense" as the idealized morphogen defined by the French Flag Model, and the work by Jaeger et al. as doing away with the idealization (5), as well as (2), (7) and (8).[17] Thus, time cannot be abstracted (contra assumption 2), precise compartment boundaries do not result from morphogen thresholds (contra 5), variation in morphogen cannot fully account for differences in cell fate (contra 7), and the "generation" and "interpretation" of positional information are not separate processes (contra 8).[18]

---

[17] It should not be forgotten that, of course, dynamical models such as gene circuit models introduce their own idealizations, which are distinct from those of the French Flag Model. In general, there may not be such a thing as idealization-free representation.

[18] We could read these somewhat metaphorical expressions as referring to the formation of a set of morphogen gradients and its causing differences in gene expression, respectively.

These findings were confirmed by numerous other studies reviewed in Briscoe and Small (2015) and summarized by these authors as follows: "morphogens provide asymmetry but not precise positional information."

Another idealization concerns the absence of noise in Wolpert's model. Of course, this assumption is false, and morphogen gradients exhibit considerable random variation between individual embryos. This raises the obvious question whether the correlation between position and the expression levels of certain genes are strong enough to give each cell along an embryonic axis a unique identity. Using mathematical information theory, Dubuis et al. (2013) investigated this question by using the *Drosophila* gap gene system. They found that the positional information (understood as the mutual information of gene expression level and position) contained in the expression levels of four gap genes is sufficient to determine position along the AP axis with 1% precision (Love, 2020b). Thus, contra (6), noise can be taken into account, and the idealization (4) of deterministic causality can be relaxed.

The last de-idealization to be briefly discussed concerns assumption (9), which limits the causal influence of morphogens to their concentration. There is evidence to show that this isn't the only relevant variable; the response to the morphogen may also depend on the total time during which a cell was exposed to the morphogen (Pagès and Kerridge, 2000).

These findings make it rather difficult to say what the causal role of morphogens really is, apart from the fact that their concentration is causally relevant in the complex set of processes that generate axial embryonic patterns. Maybe there is no more than a weak family resemblance between the current dynamical models and the simple French Flag Model. I would even go as far as to suggest that the very *concept* of a morphogen as it is contained in the French Flag Model is a useful idealization that has guided research and helped scientists to formulate research questions rather than a fundamental explanatory principle. Many scientific publications in this area begin with a reference to Wolpert's theory (which is not identical with the French Flag Model; see Section 3.1) and then subject one or several of the idealized assumptions of the French Flag Model to critical scrutiny, often using mathematical models as well as experiments. This suggests that the role of the idealized model and perhaps even of the concept of morphogen itself is to help scientists make predictions, formulate specific research questions and test specific hypotheses about causality in developmental systems. Eventually, the idealizing assumptions of the initial model were replaced by more realistic accounts of morphogen action (to be continued).

The usefulness of idealized models in orienting research has been noted by other philosophers of science, in particular Wimsatt (2007: Chapter 6). The same is true for the use of scientific concepts as tools (Feest, 2010). Much philosophical work on idealization has focused on mathematical models and computer simulations. What the case of the morphogen as conceptualized in the French Flag Model teaches us is that even a *qualitative* causal model can be at the same time highly idealized and useful as a research tool. Idealization is usually thought to be mainly a feature of mathematical models. Furthermore, some distinctions within causality such as Waters's (2007) notion of actual-difference making cause turn out to be useful idealizations in some cases.

3.4 Morphogenetic Fields and Self-organizing Gradients: Driesch's and Spemann's Puzzles Solved?

During the 1990s, research on morphogen systems has led to a resurrection of a classic notion in developmental biology, that of a morphogenetic field (see Section 2.1).[19] Recent formulations of the concept of morphogenetic field have emphasized that "embryonic regions with equivalent developmental potential, or morphogenetic fields, have the remarkable property of 'regulating' to re-form a normal structure after experimental perturbations" (Reversade and De Robertis, 2005: 1147). Thus, the pattern-forming systems operative in embryo-genesis have some robustness properties that gradient systems could potentially explain. In addition, there is still the question of the scale invariance of pattern formation, as Wolpert formulated it already in his famous French Flag Problem (see Section 3.1). The scale invariance doesn't come for free with the gradient model (in fact, some have used it as evidence against the gradient hypothesis); its explanation required sophisticated systems-biological approaches.

For example, Ben-Zvi et al. (2008) developed a dynamical model for the *Xenopus* dorsoventral gradient system (the system depicted in Box 1). The model consists of a system of differential equations containing variables that represent the concentrations of several gradient-forming signaling proteins. There was already some evidence that a number of signaling proteins including some BMPs and Chordin form a self-regulating dynamic activity gradient with positive feedback along the dorsoventral axis in *Xenopus*, stretching from the organizer to a ventral signaling center at the opposite end. "Self-regulating" means that the gradient re-forms at the right scale after experimental perturb-ation (e.g., in half-embryos). This was highly suggestive that it somehow was the basis of the morphogenetic field along the D-V axis. Understanding how this was possible requires a quantitative dynamical model using differential equa-tions, which is what Ben-Zvi et al. (2008) provided. The problem was that they

---

[19] The concept has, of course, changed since its first inception. An important difference is that fields are today conceived as having well-defined boundaries, which is not the case according to earlier formulations (Davidson, 1993).

knew the components that were involved in the self-regulating gradient (the ones depicted in Box 1), but not their properties such as diffusion rates and binding affinities of the various molecular interactions. Thus, they were able to write down differential equations for the concentrations of the diverse proteins, but they did not know what numbers to plug into the model's parameters. Thus, they adopted a brute-force approach in order to screen for combinations of parameter values that would generate the right kind of behavior, namely the scaling behavior.

The basic model contains nine parameters, including the diffusion coefficients for the different proteins and various kinetic and binding constants. The model's variables include the concentrations of the various components as a function of time, and a combined signaling rate for the BMPs. Importantly, the model contains a term for "shuttling," which is the transport constant for a complex of the BMPs with its antagonist Chordin. Using these equations, Ben-Zvi et al. then ran numerical simulations for 26,000 different parameter combinations in order to identify those combinations under which the system showed the scaling properties to be explained. Only 21 such combinations predicted such a behavior. What they found was that scaling is possible under the condition that some proteins use others as a shuttle to move along the D-V axis; the BMPs move through the embryo much more rapidly when bound to their inhibitor, Chordin. Thus, when Chordin "shuttles" its ligands effectively, the BMP signaling gradient exhibited the necessary scaling properties (see Figure 4). What this means is that the signaling activity of the gradient showed roughly the same level at any given fraction of the embryo's height irrespective of the embryo's size, and that this activity profile was restored after experimental perturbations such as cutting the embryo in half.

Of course, the model contains various idealizations. In particular, it included just a single spatial axis while, of course, real embryos are three-dimensional. This was reasonable given the properties of such gradients. Even if the DV-axis doesn't have the same length at all cross sections, this doesn't seem to matter if the gradient is size-invariant. As is always the case, there were other well-justified idealizations. Given our discussion in Section 3.2, perhaps the most significant idealization is the assumption that one of the proteins, presumably Chordin (see below), will act as a *full* morphogen in the sense explicated there. At any rate, I would like to work out a different feature of the model here, namely its *how-possibly* character.

Philosophers of biology have pointed out that some models or explanations do not describe the *actual* mechanisms responsible for a biological phenomenon but merely the *possible* mechanisms (Brandon, 1990). According to this

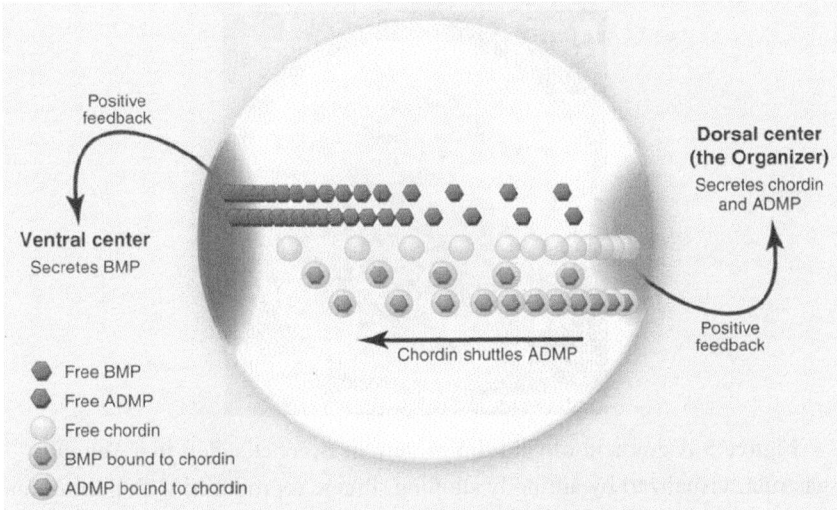

Figure 4 The model of self-regulation of the BMP/Chordin signaling gradient (see also Box 1). A combination of positive feedback (proteins stimulating their own production), inhibition (of BMP and ADMP activity by Chordin) and transport or "shuttling" of ADMP by Chordin creates a dynamic equilibrium that generates the same activation pattern of BMP irrespective of size. Image reproduced with permission from Lewis (2008).

distinction, how-possibly explanations have all the needed explanatory virtues (which depend on the investigative context and the exact questions asked), except that there is no or insufficient evidence to consider it as the actual explanation. How-possibly explanations often come in groups, that is, there are several candidate explanations or models that would provide an explanation if they were true.

I suggest that the model of the BMP-Chordin self-regulating morphogen gradient presented earlier provides a how-possibly explanation. The reason is that many of the parameter values of the model remain unknown, even if some of the predictions of the model were confirmed experimentally; for example, that the BMPs are shuttled away from their site of production. However, there are several parameter combinations that are consistent with the experimentally ascertained facts. Thus, it is only known what parameter combinations *would* allow for scaling and robustness of the gradient, not which combinations are *actually* responsible for these effects. Lewis (2008: 401) writes that the model "highlights the central frustration of mathematical modeling in developmental biology: For most of the signaling systems that pattern the embryo, with their intricate feedback loops, we simply do not have

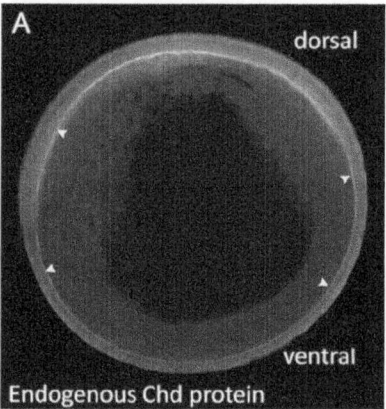

Figure 5 A gradient of Chordin protein in Brachet's cleft in a *Xenopus* gastrula, visualized by antibody staining. Image reproduced with permission from Plouhinec et al. (2013).

the quantitative data that are needed to go beyond proof of plausibility to the type of solidly based quantitative theory that is commonplace in the physical sciences."

While these uncertainties remain, a crucial role for Chordin is supported by a spectacular study in Edward De Robertis' lab (Plouhinec et al., 2013). They were able to visualize the Chordin gradient in Brachet's cleft, a thin layer separating ectoderm and mesoderm in the amphibian gastrula (see Figure 5). Presumably, the extracellular matrix in Brachet's cleft provides favorable conditions for the shuttling of Chordin. What is even more spectacular, when they cut the embryos in half such as to create a dorsal and a ventral half, the Chordin gradient re-formed to scale in the dorsal half. Indeed, the dorsal half is able to form a smaller embryo, while the ventral part only forms a *"Bauchstück"* (Spemann's term) or "belly piece."

Most remarkably, De Robertis' lab showed that embryos receiving a classical Spemann–Mangold organizer transplant formed a second Chordin gradient. This could at least explain how the organizer can impose a dorsoventral pattern into the recipient embryonic tissue. The anteroposterior patterning is believed to originate in a gradient of Wnt, a signal-transducing protein that was first identified as playing a role in carcinogenesis.[20]

---

[20] Several of these proteins that are active early in morphogenesis keep important jobs in the adult organism, where they function in various signal-transducing pathways that regulate cell proliferation and hormonal regulation.

Thus, it seems that there are at least how-possibly solutions for Driesch's and Spemann's puzzles about the remarkable self-regulating powers of morphogenetic fields. The robustness and scaling properties of morphogenetic fields can be accounted for by the self-regulating morphogen gradients that cause the first anteroposterior and dorsoventral patterns. This self-regulating property can be understood by the complex dynamics of the morphogen system involving various kinds of feedback.

Thus, the robustness as well as scaling properties of morphogen gradients can be accounted for by dynamical models with differential equations, which can be solved by numerical simulations on a computer. What these simulations provide are steady-state conditions, that is, combinations of values for the different variables under which there is no change and therefore a stable gradient. The steady state is dynamical, which means that processes such as diffusion and chemical reactions such as cleavage of Chordin by its protease still occur, but the net values of all the variables stay the same. This raises the question if such a dynamical account is compatible with the causal perspective.

Various concerns have been raised in the philosophy of causality and philosophy of systems biology literatures, some of which we have already briefly touched upon in Section 1.5. One concern is the extent to which causal representations that essentially contain difference-making information can be integrated with the dynamical information provided by differential equation models (Woodward, 2013). This kind of integration is possible by interpreting the differential equations themselves as a structural causal model (Anderson, 2020; Meyer, 2020; Weber, 2016).

Another concern is whether equilibrium explanations in general are causal at all. The reason is that such accounts basically give the conditions for equilibrium or steady state, while abstracting from any specific causal paths at which the steady state is reached. However, such equilibrium explanations fit well into the interventionist causal perspective outlined in Section 1.5.

At any rate, the most significant findings from this analysis are (1) the *abstract* and *idealized* nature and (2) the *how-possibly* character of the existing model-based explanations of the phenomena associated with morphogenetic fields. Thus, the epistemic conquest of such an intricate causal system comes at a price. Idealization, abstraction and causal selection to guide abstraction are indispensable for understanding such a vastly complex system as a developing embryo. However, it should be noted that how-possibly explanations such as the one that we have considered in this section are sufficient for undermining Driesch's central argument for vitalism (Section 1.1); it doesn't take how-actually explanations. For in order to disprove the claim that mechanistic explanation of morphogenetic field phenomena is impossible, it suffices to demonstrate the existence of a *possible* mechanism that can account for these phenomena, that is, a mechanism that is compatible with the known experimental facts as well as with accepted physical-chemical principles, which is the case with the BMP-Chordin self-regulating gradient account of organizer phenomena in vertebrates.

Finally, my analysis in this section and Section 4 shows that all it takes to understand these properties are dynamical causal models. It should also be noted that none of these causal notions involve macro-determination or top-down causality, which would disappoint Driesch and his followers with their *Ganzheitsfaktoren*. Morphogenetic fields are no more holistic or emergent than other dynamical systems showing stable equilibria (if emergence is understood as inexplainability from the properties of the parts; see Section 1.2).

4 Causes in Harmony

In the previous two sections, we have seen that developmental biology attempts to understand the enormously complex processes occurring during embryonic development by focusing on specific types of causes, namely causes that are specific either in the sense of binding specificity or in the sense of fine-grained control (Section 2), and also by using idealizations (Section 3). But there is more. In this section, I will examine some additional types of causes that are of particular interest for developmental biologists, namely the concepts of instructive versus permissive cause (Section 4.1), signal (Section 4.2) and selector gene (Section 4.3). Finally, I show in Section 4.4 that all these kinds of causes have something in common, something I will call *causally coherent control*.

4.1 Instructive versus Permissive: Developmental Biology's Own Distinction within Causality

As I already mentioned in Section 2.1, developmental biologists have a distinction within causality of their own, that between instructive and permissive causes of development. To avoid confusion, it should be mentioned at the outset that the notion of instructive cause has nothing to do with the popular metaphor of "genetic instruction." Woodward (2010) has tentatively suggested that the instructive/permissive distinction is related to what he calls causal specificity (in the fine-grained sense; see Section 2.3). In this section, I shall provide a causal analysis of this distinction. As I will show, it cannot be adequately captured by using the concept of causal specificity in the sense of fine-grained control.[21] Rather, we should view it as an irreducible distinction that is tailored precisely to the needs of developmental biology. Developmental biologists draw their own distinctions within causality that do not necessarily match these drawn by philosophers of causality.

Wessells (1977) has proposed three criteria for *instructive* tissue interactions (46–47), which I have somewhat amended:

(1) in the presence of tissue A, the responding tissue B develops in specific way W_B

(2) in the absence of A and potential nonspecific stimuli,[22] tissue B fails to develop in way W_B

(3) in the presence of A, tissue C, normally destined to mature in specific way W_C, is altered to develop in way W_B

Wessells considers a fourth criterion, having to do with specificity:

(4) The responding tissue should not develop in the specific way in response to nonspecific stimuli

However, this fourth criterion is not really endorsed by Wessells; he just proposes to "keep it in mind." Other biologists explicitly reject this criterion as being required from instructive causes, for example, Slack (1993).

Having thus defined instructive interactions, Wessell defines *permissive* interactions as any interactions satisfying (1) and (2), and perhaps (4), but not (3). He writes that (3) is "the only one that demands *control of selectivity in gene usage*" (46). Furthermore, he affirms that the responding tissue must be "competent" to respond (46), indicating that additional concepts are needed to understand instructive interactions.

As an example of an instructive interaction, Wessells presents the case of lens induction by underlying optic vesicles (see also Section 2.1). For the permissive interactions, he gives the example of embryonic mesoderm cells, which support mitosis in embryonic epithelial cells. None of these interactions are specific, as lenses can be induced without underlying optic vesicles ("free lenses," see

---

[21] Calcott (2017) also argues for this conclusion and provides an elaborate alternative analysis based on Waddington's well-known idea of epigenetic landscape. Bourrat (2019) defends an analysis of the instructive/permissive distinction in terms of causal specificity, assuming that it has something to do with the distinction between background and triggering conditions. Using mathematical information theory, these accounts are quite technical and thus cannot be given an adequate treatment here.

[22] The condition about nonspecific stimuli in (2) is not in Wessells's original account, but I think it is needed if condition (4) is not considered as necessary. I wish to thank an anonymous reviewer for pointing out this difficulty.

Section 2.1), and embryonic epithelial cells can grow in the presence of cells other than embryonic mesoderm cells.

I shall now proceed to an analysis of the distinction as formulated by Wessells. The first thing to note is that criteria (1) and (2) are basic causal criteria. While Wessells doesn't use counterfactual conditionals (which most philosophers think are required for causality) but states a mere correlation, I think it is clear from the context that Wessells isn't talking only about a correlation. The example makes this clear: Experiments have shown that the optic vesicle *causes* a lens to appear in ectoderm cells of different type, including cells normally destined to form epidermis.[23]

What distinguishes instructive from permissive causes on this account is that only the former are capable of somehow enabling a choice between *alternative* developmental pathways. This is why Wessells says that only instructive interactions require "control of the selectivity of gene usage." This is also in line with a characterization of the instructive/permissive distinction due to Slack (1993). Slack explains:

> In the case of a permissive induction there is only one possible outcome: in the presence of the signal the cells proceed along their normal pathway, in its absence their development is arrested and they fail to differentiate (Slack, 1993: 91).
>
> (. . .)
>
> In the case of an instructive induction there are at least two possible outcomes. One is autonomously achieved in the absence of the signal, the other is achieved in its presence (92).

Slack also claims that instructive induction always leads to an "increase in complexity of the responding tissue" (92). I think it is obvious that by "only one possible outcome" in the permissive case, Slack means: only one developmental path that can go forward. (Note that we could also count the outcomes by saying that there are two outcomes: either development goes forward or it is arrested.) In the instructive case, by contrast, Slack has two possible outcomes, which means two different ways in which development could go forward. For example, ectoderm cells could go forward to form epidermis or they could form a lens. A developmental arrest is not a possible outcome here. If the instructive signal is there, a lens will form, otherwise epidermis (except if there is a nonspecific stimulus).

---

[23] The causal condition could also be stated in this way: the tissue A is a necessary part of a complex of causes that includes the responsive, competent tissue and that is sufficient for the responsive tissue to develop in way W_B. But the whole complex is only necessary in some circumstances, in particular in the absence of nonspecific stimuli.

Instructive causes come in two types on Slack's account: appositional and gradient-like. The lens induction case is of the appositional type, where there are two possible outcomes (lens induction or epidermis). In the case of morphogen gradients (see Section 3), there are more than two outcomes, depending on the morphogen concentration.

Slack writes: "the important thing is the number of choices represented by the competence of the responding tissue" (92). This could be taken to indicate that, as some philosophers have suggested, the instructive/permissive distinction is a matter of causal specificity. However, I will show now that this fails to capture the nature of this distinction. Instructiveness is not a matter of causal specificity in Woodward's sense at all, for the distinction between permissive and instructive causes is a qualitative one. The nature of instructive causes is that they control the choice between *alternative developmental pathways*, while permissive causes merely control whether or not a *single* pathway goes ahead or not. Furthermore, the nature of the signal in the instructive case is such that the system follows a *default* pathway absent the signal, and an *induced* pathway in its presence. In the permissive case, the default is arrest. I suggest, therefore, that we are dealing here with an altogether different distinction within causality than any of the ones that have previously been discussed in the causation literature. It is developmental biology's own distinction within causality. Furthermore, this distinction cannot be captured in purely abstract causal terms; we can only understand it by appealing to the biological concept of developmental pathway or fate.[24]

Three further characteristics of permissive and instructive causes are worth mentioning. First, instructive as well as permissive causes are usually understood as acting *cell-non-autonomously*. This means that such a cause can act in other cells than the one that has produced the instructive or permissive signal. By contrast, cell-autonomous causes only have an effect on the cells that produce this cause. An example is the selector genes (see Section 4.3). Second, Slack's distinction between appositional instructive cause and morphogen is a matter of causal specificity. So, although causal specificity does not suffice for distinguishing instructive and permissive causes, there are differences in causal specificity *within* the class of instructive causes. As we have seen (Section 2.3), causal specificity also characterizes the Spemann–Mangold organizer to some extent, which also qualifies as instructive cause. Thus, Woodward's suggestion was not completely off the track. Third, I think we can only fully understand the notion of instructive cause on the grounds of the notion of coherent causal control, to be elaborated in Section 4.4. For what

---

[24] Thanks to Ulrich Stegmann for pointing this out to me.

characterizes instructive as opposed to permissive causes is their effect on gene expression in the responsive tissues. But the idea is not simply that such causes affect the activity of numerous genes. Certain drugs that are not understood as instructive causes could have the same effect. Rather, instructive causes control their target genes in a *coherent* fashion, allowing them to adopt some specific developmental fate.

4.2 Signaling Pathways, Quasi-Conventionality, and Modularity

The concept of signal – not used by classical experimental embryologists – is ubiquitous in molecular biology. Practically all known cells, including even simple bacteria, are said to exchange chemical signals with each other. Such signals allow them to coordinate their growth as well as their performance of physiological functions. Hormones such as insulin or estrogen are also described as signals. We have seen in Sections 2 and 3 that molecules such as the bone morphogenetic proteins and other gradient-forming morphogens are considered to be parts of signaling pathways involved in determining the developmental fate of embryonic cells in the function of their position. Box 2 shows such a pathway.

Many philosophers of science would say that what Box 2 represents is simply a mechanism (see also Section 1.3). While this is certainly the case according to most accounts of mechanism, we must be careful not to obscure important differences between different kinds of scientific explanations by subsuming everything under the mechanism concept (Ross, 2020). Ross has proposed that explanations invoking pathways in biology have the following features: They (i) capture sequences of steps, where these steps (ii) track the flow of some entity or signal through a system, (iii) abstract from significant causal detail, and (iv) emphasize the "connection" aspect of causal relationships (Ross, 2020). Some of these features may be found in mechanistic explanations as well, in particular the sequential (i) and connection (iv) aspects. By contrast, the flow (ii) and abstraction (iii) aspects are not usually found in mechanisms.[25]

Now, it seems to me that signal transducing pathways have some of these features; however, I think the signal-tracking aspect needs explication. What does it mean that such a pathway tracks a signal? It seems that what is being transmitted is an *activation state* or sometimes an *activation level* (in the case of graded responses, e.g., to morphogen concentrations).[26] For example, in the

---

[25] It should be noted that some proponents of New Mechanism take a nuanced view and argue that mechanistic explanations can admit abstraction, for example, Craver and Kaplan (2020).

[26] Some people would say that what flows through a signaling pathway is information, but this notion needs unpacking. I lack the space here to analyze this notion. Mike Stuart tells me that, in his interviews, systems biologists tend to treat the transmission of information and of activation

Box 2 EXAMPLE OF A SIGNALING PATHWAY USED IN EMBRYONIC INDUCTION

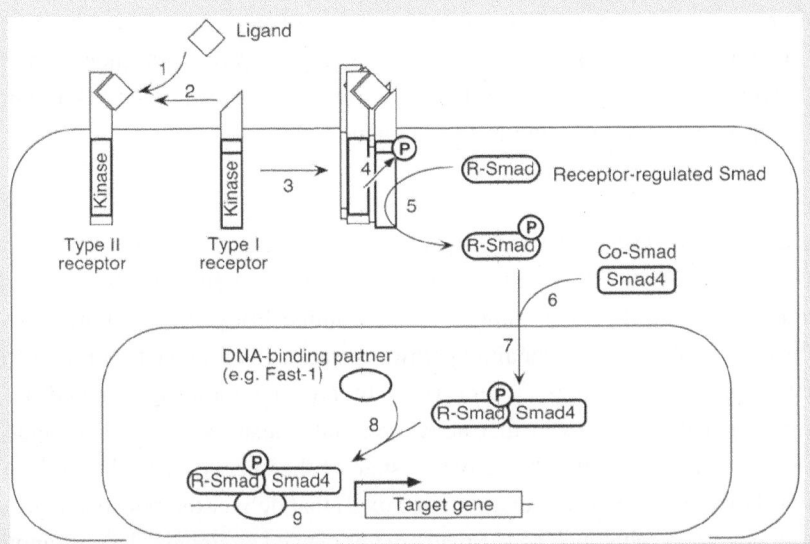

This diagram shows the pathway associated with the family of TGF-beta (transforming growth factor) signaling proteins. In this family we find, among others, the bone morphogenetic proteins (BMPs) responsible for some of the effects of the Spemann–Mangold organizer (see Box 1). The BMPs were initially discovered by their power to stimulate the growth of cultured bone cells; thus, like many signaling molecules, they have more than one biological function. The extracellular molecule labeled "ligand" in this diagram could be such a BMP, but there are numerous other possible signaling molecules in this class. Like many other similar pathways, the TGF-beta signaling pathway can function in the regulation of a variety of biological processes, depending on its cellular context as well as on the developmental stage. Upon binding to cell surface receptors, the ligand forms a complex with type I and type II receptors, which leads to the attachment of a phosphate group to receptor type I. (A kinase is an enzyme that attaches phosphate groups to other proteins.) Thus activated, the receptor-ligand complex phosphorylates a molecule of the receptor-regulated Smad (R-Smad) family (named after the *Drosophila* gene "*Mad*" where it was first discovered). This allows the R-Smad protein to form a complex with another protein from the same family, Smad4. This complex is then transported into the cell nucleus, where it binds the DNA-binding protein Fast-1 ("*forkhead* activin signal transducer"). The whole complex then binds to DNA at some specific site, thus activating one or several target genes. In the signaling pathways involved in developmental processes, there are usually a large number of target genes, as they are involved in the control of extremely complex processes such as the formation of neural tissue from yet undifferentiated cells. Diagram reprinted with permission from Massagué (1998).

example of TFG-β signaling, ligand binding activates the receptor kinase, which activates the R-Smad kinase, and so on (see Box 2). It's a bit like dominos, where the state "down" of the domino bricks flows along a track. Thus, we could say that signaling pathways track activation states or activation levels through a complex system. But why are such pathways singled out as *signaling* pathways, given that activation states also exist in other kinds of pathways? For example, the activated state of the energy-rich ATP molecules (not shown in the pathway diagram in Box 2) that deliver the phosphate groups to signaling proteins such as R-Smad are not viewed as signals. To give another example, plant chlorophyll molecules can transmit an excited energy state to neighboring molecules, which is not described as a case of signaling. So what makes a signaling pathway?

A potential distinguishing feature of signaling pathways that may be absent in other kinds of pathways is their *quasi-conventionality*. In human communication, signals such as words or gestures like "thumbs up" normally do not come with a natural meaning; it requires a convention between the users of the signals. Philosophers have used game theory to show how conventions can arise between rational players even in the absence of explicit agreement. This idea has also been applied to the evolution of signaling systems in living organisms (Skyrms, 2010). Thus, conventionality could be a defining feature of signaling systems even outside the realm of human communication. This raises the question of whether the molecular signals that, as we have seen in previous sections, play such a crucial role in controlling development also manifest this feature of conventionality or if it is even among their essential features.

Skyrms (2010) has looked at the example of quorum sensing in bacteria. This is an interesting phenomenon that might well be the evolutionary origin of all intercellular signaling. Some bacteria produce and secrete substances called "autoinducers" that allow them and other bacterial cells to "sense" the

---

states as being one and the same thing. Biologists also refer to the transmission of activation states as "signal transduction," suggesting that they consider it as a distinct kind of process.

population density of cells of their own or of another species. The more bacteria there are in the vicinity, the higher the concentration of this molecule will be. The cell will only activate certain genes if the concentration is below or above a certain threshold. Bacteria use this system to regulate not only their growth but also activities that are beneficial only when enough bacteria are engaging in them together, such as bioluminescence, sticking together to form biofilms, or antibiotic production to fight off competitors.

Intuitively, it makes sense to say that the autoinducer signals are conventional, because the chemical composition of the autoinducer does not matter at all (it can be a polypeptide or some small molecule) as long as the bacterial cells can recognize the signals and mount a response to them. Nonetheless, Skyrms (2010) shies away from attributing full-blown conventionality to autoinducing signals: "Pure convention is gone, but development of the same ancestral signaling system could go in one way or another—and in different species of bacteria has done so" (31). Furthermore, Skyrms suggests that there could be "degrees of conventionality associated with degrees of plasticity in signaling" (31).

The question is if our causal perspective is sensitive to this quasi-conventional aspect of signals. After all, it seems that signals are just causal difference-makers; so how can we capture this aspect? And how could we make sense of Skyrms's idea that there are "degrees of conventionality associated with degrees of plasticity" ?

One idea might be to understand the notion that "it doesn't matter what the signal is" in terms of *abstraction*. This would move signaling pathways closer to Ross's account of pathways briefly presented above. The idea that a description in terms of signaling systems involves a certain amount of abstraction from causal and/or mechanistic detail has been ably defended by Levy (2011). However, if we follow Skyrms, the point about conventionality has to do with the existence of *alternative material realizations* of the signaling interaction. When we confer the status of signal to a biochemical entity, we want to attribute to them what metaphysicians call a *modal* property: the property that, essentially, a signal's chemical realizer *could* be different from what it actually is. As Skyrms writes: "Conventionality enters when there is enough plasticity in the signaling interactions to allow alternative signaling systems" (2010: 31). The problem is how to understand the nature of this "plasticity."

Perhaps one way to understand plasticity and, hence, conventionality in causal terms is by using the notion of *modularity*. A causal structure is modular to the extent to which it is possible to change some of its causal relationships (e.g., a component or a set of components), while leaving other relationships intact. A simple example discussed by Woodward (2013) is a causal structure in which a match is struck, which ignites a fire, which boils a kettle of water. Parts of this causal structure remain intact even when the match is soaked. The fire that heats the water could be started by something other than the match. Woodward thinks that modularity is a general feature of mechanisms, thus, the notion may also apply to many pathways and hence not be that useful in distinguishing signaling systems from other causal systems.

Perhaps, when we look at a typical signaling pathway such as the TGF-β pathway shown in Box 2, what is striking is the high degree of modularity: the entire pathway, including the first (i.e., the extracellular ligand) and the last component (the transcription factor/s), could be replaced by a different one so long as the first component in the pathway retains its binding specificities on the input side, and the last component on the output side. If this were done consistently in an organism in a way that doesn't interfere with other signaling systems, everything else could function just the same. I am not sure if this is true of many other causal pathways. For example, in a metabolic pathway at least the first and last compound must be identical for this pathway to be substitutable by a different pathway. In a signaling pathway, by contrast, it is enough if the first and last components have the same binding specificity; they don't have to be identical in any other respects. Modularity is *complete*.

Nonetheless, modularity is hardly sufficient to distinguish signaling from more ordinary causal pathways.[27] Many mechanisms are modular, too. For this reason, I suggest that the reason why such pathways as the one discussed here are singled out as *signaling* pathways is their involvement in what I call *coherent causal control*. This idea will be presented in Section 4.4.

4.3 Selector Genes

Another important causal notion in molecular developmental biology is that of selector genes. Actually, this concept came out of classical genetic studies: García-Bellido (1975) grouped all genes controlling development into either "cyto-differentiation" or "selector genes" (161). The former group includes genes the products of which are involved in controlling cell division and other behaviors of the individual cells. They are also called "realizator genes." The latter group are those that "control developmental pathways." According to García-Bellido, they are "characterized by the fact that they change the overall

---

[27] Artiga (2020) has suggested that signals are *minimal* causes, i.e., individual causal factors that are causally relevant without constituting an enabling mechanism. This doesn't seem to be sufficient, as, for example, the ATP molecules that deliver phosphate groups in the pathway in Box 2 satisfy this criterion without being themselves signals.

organization of a developmental system without affecting normal cytodifferentiation mechanisms" (1975: 169). This concept was widely adopted, and different types of selector genes are recognized today: field-, region-, cell-type, tissue- and compartment-specific (Mann and Carroll, 2002). As their name suggests, these genes select in some defined area of the embryo which other genes are activated. Their effect is that one among several different developmental pathways is selected. All selector genes are transcription factors that directly bind to DNA, often together with several cofactors, to regulate the expression of nearby target genes. Gradient-forming morphogens (see Section 3) determine what selector genes are activated when and where in the embryo. In contrast to the morphogens, selector genes act cell-autonomously, which means that they only have an effect on cells which express them (see Section 4.1). Thus, selector genes are parts of intricate networks of transcription factors that are said to "interpret" the positional information provided by morphogen gradients.

While the concept of selector genes came from classical genetic research on wing development in *Drosophila*, several other classes of genes were eventually admitted to the category of selector genes. I shall briefly discuss two examples: the Hox genes, a class of region-specific selector genes, and *eyeless/Pax6*, a field-specific selector gene. Hox genes are involved in patterning the embryo along the anteroposterior axis. They determine the identity of each one of the segments on an insect embryo (see Box 3).

It has not been easy to identify the genes on which the Hox genes act. As usual, the best understood systems are found in *Drosophila*. For example, Lovegrove et al. (2006) have studied the coordinated control of several target genes of the Hox-gene Abdominal-B during development of the respiratory organ of the larva, known as the spiracle. Abd-B (see Box 3) specifies the identity (i.e., normal development) of the last five posterior segments of the fly. According to the study by Lovegrove et al., Abd-A first activates a layer of four regulator genes, each of which regulates the activity of a set of realizator genes belonging to three functional groups, namely genes encoding cell adhesion proteins (molecules that make specific cells stick together), cell polarity genes (encoding proteins differentiating the cell membrane into a side facing the surface and one facing the interior of the tissue), and proteins regulating cytoskeleton formation (cytoskeleton: intracellular protein fibers that give the cell stability as well as allowing cell movements). Abd-B and the four regulatory genes thus coordinate the activities of these genes, determining the cells to form specific structures such as the spiracles. Figure 6 shows coordinated control of gene expression by the Hox gene Abdominal-B in the fruit fly.

Box 3 Hox Genes

Hox genes owe their name to a DNA sequence element known as the "homeobox," which they all contain, even though they are not the only genes with this sequence element. The Hox gene family is characterized by their role in specifying the body plan along the anteroposterior axis in most animals including humans. The name "homeo" (from the Greek word for "the same") stems from a class of mutations that have been named "homeotic" by classical geneticists, who knew these mutants for a long time, the first ones being described as early as 1915 in *Drosophila* (of course). In classic homeotic mutants, a segment of the fly body is transformed into a segment that belongs before or after that segment. For example, in the naturally occurring *Drosophila* mutant *Bithorax*, the third thoracic segment is transformed into a second segment carrying an extra pair of wings. A remarkable feature of Hox genes is the fact that they act combinatorially. This means that a segment will assume a specific identity as a function of what combination of Hox genes is active in the cells. What combination is active is directly determined by early morphogen gradients.

The figure (modified from Wikimedia Commons File:Hoxgenesoffruitfly.svg) shows the two Hox gene complexes of *Drosophila*, named Antennapedia (ANT) and Bithorax (BX) complex. Interestingly, the eight Hox genes of these complexes are arranged on the chromosome in the same order as they are expressed along the animal's main axis.

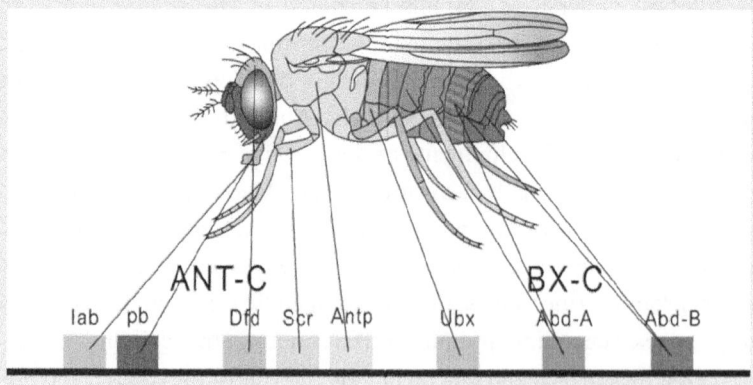

Another example of a selector gene is the *Drosophila* gene *eyeless*, which is homologous to the gene *Pax6* in vertebrates (including humans). This gene seems to be able to initiate eye development and is classified as a field-specific selector, because the eye forms a classic morphogenetic field (see Section 3.5).

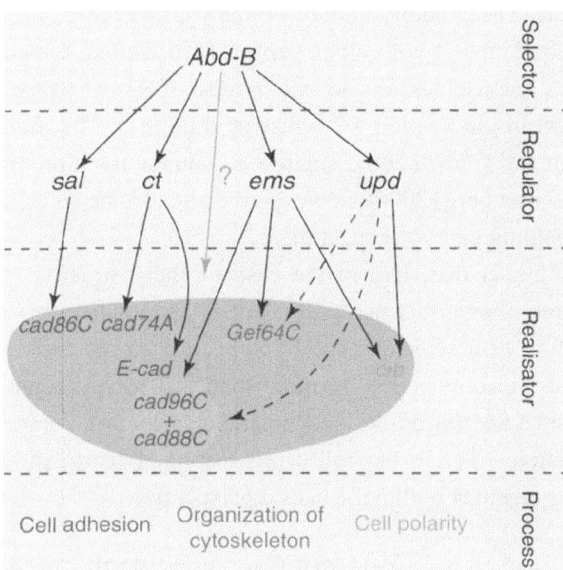

Figure 6 Coordinated control of gene expression by the Hox gene Abdominal-B in the fruit fly. Image reproduced with permission from Lohmann (2006).

This power was spectacularly demonstrated by flies that were genetically manipulated such as to express the product of the *eyeless* gene in different parts of the fly, instead of just in the head. These flies have eyes everywhere (Halder et al., 1995).

As these examples show, mutations in selector genes can have dramatic effects on an organism's body plan, which is why they are thought to play a major role in macroevolution. They are thus the bread and butter of the thriving field of developmental evolutionary biology or "evo-devo." These dramatic effects are thought to be due to the "collapse or perturbation of networks of selector-regulated genes" (Mann and Carroll, 2002: 592). This ability of collapsing or perturbing gene regulatory networks (when mutated) is indicative of the central place occupied by these selector genes within the networks. This central role has often been circumscribed by using metaphors such as "master control genes" (Gehring, 1998), but this is hardly a precise idea. How can we describe their causal role while staying true to our causal perspective?

An important insight concerning the causal role of selector genes concerns their binding sites on the DNA. These are described by Mann and Carroll (2002: 592) as "cis-regulatory DNA elements that act as developmental integrators of selector protein function." By "cis-regulatory DNA elements" they mean sequences that are located near a gene that is regulated by the selector gene. The reason for the selector genes' central role thus has to do with the fact that they bind to the cis-regulatory regions of many target genes and that they do so typically in combination with other transcription factors, which can be either selector genes themselves or the end points of signaling pathways as we discussed them in the Section 4.2 (Guss et al., 2001). The term "integrators" used by Mann and Carroll could suggest a computational reading such as the one given by Rosenberg (2006); however, it could also be taken to simply mean that many signaling pathways meet there.

My contention is that, like in the case of the instructive signals (which includes the morphogen discussed in detail in Section 3), we can only understand the notion of a selector gene by appealing to the concept of coherent causal control. Selector genes are more than just highly connected nodes in complex causal networks or top-level positions in control hierarchies; they take on a coordinative role. In the following section, I try to provide a way of understanding this idea within the causal perspective.

4.4 Coherent Causal Control

In the previous three sections I have discussed some causal notions that are frequently encountered in developmental biology, namely the concepts of instructive and permissive cause, signal and selector gene. I would now like to suggest that these causal notions share an important property, which, to my knowledge, has not yet been noticed by philosophers.[28] The reason why they haven't been noticed may have to do with contemporary accounts of causality, which have focused mainly on the *structure* of causal networks, which can be represented by directed graphs (Woodward, 2003). Causal graphs normally represent causal dependencies between causal variables that can take different values. Let's consider a very simple causal structure: a dry match that is struck in the presence of oxygen will ignite a fire. The corresponding causal graph will consist of three nodes and three arrows pointing to a fourth node, one representing the dry match, one the striking, one the oxygen, and one the fire. The nodes represent variables that can take the values 0 or 1, where the 0 stands for the absence and the 1 for the presence of the factor. The effect, too, can be represented by 0 or 1 in this case, as it may occur or not occur. In this graph, the causes interact in the sense that all the three causes have to be present for the effect to occur. This interaction is not reflected in the structure of the graph, but it can be captured by a type of mathematical equations called structural equations.

---

[28] Perhaps the closest existing idea is Driesch's notion of *Kausalharmonie* (Driesch, 1905: 175). There might also be some resemblance to Calcott's (2017) hierarchical analysis of interacting causes or to Bich et al's (2016) account of biological regulation, but I lack the space to explore the possible relations to these accounts here.

Now let us consider a more complex causal graph network, which has many causes and many effects; for example, the gene regulatory network shown in Figure 6. We assume that some of the variables may take more than just two possible values (i.e., they express the rates at which some process takes place). Typical signaling pathways such as the one shown in Box 2 can also be viewed as containing such a graph, but also additional information such as interactions between causes. Now, what such a graph (not identical to the pathway diagram, because the latter contains information beyond the pure causal dependencies) does *not* express is the fact that, in a living organism, the *values* that the different variables take must somehow be tuned to each other. This means that they must take values that allow the system as a whole to perform a specific activity or function. Thus, we are not merely talking about a causal graph with a central node; the values of this control variable must also set those of the controlled variables to values that somehow *cohere* with each other. Let us see what this could mean for the four kinds of causes that we have encountered. I shall start with the notion of signal, as it is the most general one.

Signals (see Section 4.2) typically have highly diverse effects on a biological system. To use a well-understood example, when the hormone insulin is secreted into the bloodstream, it causes numerous organs in the body to take up glucose. In the liver and muscle cells, insulin activates the synthesis of glycogen, which is a storage form of sugar. In addition, insulin has numerous other effects on the body (e.g., on lipid and amino acid metabolism, blood potassium concentration). Most of these effects have the same goal: to lower the concentration of free glucose in the blood. The different effects of insulin are coherent with each other in the sense that the hormone signal doesn't induce some organs to take up sugar and others at the same time to release sugar; that would seem incoherent.

I mean by *causal coherence* the way in which the variables representing these causal factors (glucose uptake, glycogen synthesis, lipid metabolism, potassium level in the insulin example) take values that help the organism lower its blood glucose level, which is a biological activity. By the same token, I shall refer to insulin's effect on the values of all these numerous causal variables as a *causally coherent response*. Finally, I shall refer to causal variables that are able to constrain the values of their causal descendent variables (i.e., the variables that are causally downstream) in such a way as a *coherent control variable*.

A formal account of coherent causal control will have to be developed elsewhere. Briefly, I see causal coherence as a relation between the *values* of a set of effect variables that are regulated by a common control variable with respect to a biological function, process or activity that performs better under some combinations of values than under other (incoherent) combinations. Thus, causal coherence is not a purely causal property as it involves an ineliminable reference to biological functions, processes or activities. But unlike Driesch's entelechy, it is explicable in strictly physicalistic terms.

All the main causal factors that we have encountered in this section as well as in the previous ones are such coherent control variables. Take the morphogens: their presence in some concentration elicits a coherent response in a tissue in that, in each embryonic cell, numerous genes (in the hundreds) must be activated, while others are suppressed, for the cell to adopt a specific developmental pathway (e.g., the pathway that leads to brain or other neural cells). As we have seen in Section 3, it is possible to manipulate morphogen gradients by experimental intervention (or by naturally occurring mutations) such as to modify the structure of the embryo along some body axis. While such manipulations will not always produce a viable adult organism, at least they allow development to go on for some time. For this to be possible, morphogens must be able to control the expression of a large number of genes in a coherent way. This means that the same group of cells will not, for example, activate genes needed for neural development and for muscle development at the same time but only those needed for a specific developmental pathway. Furthermore, the genes needed for some pathway need to be activated at the right rate to avoid, for example, excessive cell proliferation leading to hypertrophied organs, and so on. Development is tightly coordinated.

It is not my claim that this kind of coordination is achieved by the coherent control variables alone. The coherence of the response is a property of the whole system, typically consisting of a large number of proteins and regulatory sequences on DNA.[29] What characterizes coherent control variables is the fact that they can elicit a coherent response for a range of different values; in other words, the coherence of their response is *invariant* with respect to a range of possible interventions. Invariance under some range of interventions is a universal feature of causality (Woodward, 2003), but the ability to produce an invariant *coherent* response is not. Such a coherent response, I contend, characterizes mainly living organisms as well as technological artefacts with a control architecture.

---

[29] An interesting question is if the heterologous inducers mentioned in Section 2.2 should also be counted as coherent control variables. Even though this might seem counterintuitive, my contention is that heterologous inducers can elicit causally coherent responses in the sense outlined here and therefore count as coherent control variables with respect to these systems. However, their coherent effects are mediated by other variables, so they are perhaps better viewed as intervention variables with respect to coherent control variables. In any case, causally coherent control is always a phenomenon that is exhibited by a whole causal network not just by a single cause.

What causal variables are relevant for achieving coherent control? One mechanism that might contribute to causal coherence is regulatory feedback, which is easy to capture in a causal framework.[30] This may well be the case sometimes, but I don't think it is necessary. Judging from the cases studied here, it seems that most cases of coherent control are due to the *binding specificities* and the *kind* and *magnitude* of the effects that the coherent control variable has on its diverse targets. For example, for causal coherence it matters that insulin have a stimulating effect on the synthesis of the storage carbohydrate glycogen in the liver and at the same time an inhibiting effect on the enzymes that break down glycogen into glucose. This coherence is realized by the binding specificities as well as by the kind and the strength of the effects (e.g., inhibition or activation) produced by the hormone on various signal transduction pathways. In a similar way, the binding specificities of numerous signaling molecules (e.g., BMPs, Chordin) or the products of selector genes (e.g., the transcription factors produced by the Hox genes) as well as the kind and magnitude of their effects on their targets are tuned to each other in such a way as to generate signaling levels and activity patterns of their respective targets that cause cells to adopt one developmental fate rather than another. Thus, coherent causal variables control key steps in the development of an organism, and many classical as well as molecular experiments have targeted precisely such causal variables.

5 Epilogue: Understanding, Progress and Manipulability

I have shown in the previous sections how a particular causal perspective can shed light on some aspects of scientific practice in developmental biology. The grafting experiments of classical experimental embryology, crude as they were, provided nonetheless a considerable amount of causal knowledge about inductive interactions between different embryonic tissues. I have argued that this knowledge was not only explanatory in itself; it was also of great value for identifying the first molecules involved in patterning the embryo especially in amphibians, which were less developed as tools for genetic research than *Drosophila*. Furthermore, this classical experimental research helped to shape some important causal concepts such as induction, morphogenetic field, morphogen, permissive and instructive interactions, and selector genes – concepts which survived the molecular revolution in biology. Nonetheless, some of the phenomena described by such concepts have become differently classified at the molecular level, which is why it may be more appropriate to speak of a replacement rather than a reduction of classical experimental embryology by molecular developmental biology; a replacement, however, which didn't

---

[30] Thanks to Mike Stuart for suggesting this.

preclude the techniques and some concepts from classical experimental embryology from being useful to molecular developmental biologists.

There are at least three aspects of scientific practice that we can understand particularly well from a causal perspective. First, we can give a rationale for the highly *selective* way in which scientists explain developmental phenomena, namely by abstracting away from a lot of details. As I tried to show in the preceding sections, contemporary understanding in developmental biology comes not primarily from knowing all the causes, processes or mechanisms that transform a zygote into an adult form. It is rather a matter of knowing certain select kinds of causes, namely such causes that allow experimenters, as well as nature herself, to control key developmental steps. I have suggested that these select kinds include such causes displaying specificity in one or both of the two senses of binding specificity and fine-grained specificity, as well as causes exhibiting what I call coherent causal control, which includes instructive causes (including morphogens), signals and selector genes. Furthermore, we have also seen that some causal models are highly idealized and that some provide how-possibly rather than how-actually explanations.

The partial nature of causal explanations is something that many philosophers of science have noticed when looking at sciences that deal with massively complex systems. Some philosophers have discussed this kind of selectivity under the rubrics of abstraction and idealization (Potochnik, 2017; Wimsatt, 2007), others by appealing to causal selection instead (Baxter, 2019; Franklin-Hall, 2015; Lean, 2019; Plutynski, 2018; Ross, 2018; Waters, 2007; Weber, forthcoming-a; Woodward, 2010). I have shown in this Element that some of the causal features used in causal selection are themselves idealizations in some cases, in particular those in Waters's (2007) notion of actual-difference making cause.

Developmental biology thus provides a primary example of the principle that understanding a complex system means to somehow curb the massive complexity by foregrounding some causes that have some special import for our understanding, given the specific epistemic goals of a scientific discipline, and backgrounding a lot of other causally relevant factors (but, of course, not all of them, and the backgrounding may not be permanent). While a broadly mechanistic approach is at least compatible with this practice (Craver and Kaplan, 2020), the currently popular philosophical accounts of mechanistic explanation by their own lights offer few resources for accounting for these causal selection practices and thus at least need to be supplemented by distinguishing between different kinds of causes that have specific explanatory value in a given investigative context.

In this Element, I have identified some types of causes that have received special attention in recent developmental biology. I think that one of the most striking results of this inquiry is the fact that these types of causes are at the same time highly relevant for our *understanding*[31] of developmental processes as they are *useful* as targets for interventions designed to learn more about the processes involved. Maybe that's not a coincidence; perhaps the scientific intellect generally likes causes that it can exploit for learning more.

But there is clearly more that could be said about *why* the causal notions I have examined might contribute more to the desired kind of understanding given biologists' epistemic goals than other causal factors (again, I do not claim that these "other" factors are completely irrelevant). I believe that this would require a *functional* approach, that is, an investigation of the goals and purposes of the reasoning tools in question (Woody 2015). So far, I can only speculate. We may assume that the notion of mechanism is linked to automata; we like it when work does itself with us only watching. A bit more seriously, notions having to do with control in all likelihood come from engineering. Watt's famous "governor" is the paradigm of a mechanism with regulatory feedback, and this concept has been enormously useful in biology. The notion of a genetic code was invented in research contexts in which cryptography was a major topic (Kay, 2000). The same practical contexts also gave us the concepts of information and signaling that are also used productively in developmental biology, as we have seen. This is just to give us some idea why the causal notions that I have examined here may have a special appeal to our understanding. It may have to do with their function in a society that heavily depends on technology. Aspects of our cognitive psychology may also play a role (Gärdenfors and Lombard, 2020). Finally, some causes may simply have received special attention because they are accessible with experimental methods.

The second aspect of scientific practice that a causal perspective highlights is the *discovery value* of certain kinds of causal knowledge in producing *more* causal knowledge. I have highlighted in particular the utility of specific causes in the fine-grained sense in discovering more causal factors thanks to targeted interventions. Such causes are found not only at the molecular level but also at the tissue level, as the case of the Spemann–Mangold organizer showed.

After the dawn of the molecular age, each newly described molecule was a potential resource for identifying more molecules involved in the same process. There are several reasons for this, as we have seen in Section 2. Molecular tools such as mRNA isolated from embryonic tissues or synthesized in vitro, as well as synthetic nucleic acids or monoclonal antibodies recognizing

---

[31] Some accounts tie scientific understanding directly to manipulability (e.g., Wilkenfeld, 2013).

specific proteins, provided tools for causal interventions that were helpful in elucidating the causal roles played by various other molecules that specifically interact with these molecular tools. This is one reason why specific causes (in the sense of binding specificity; see Section 2.3) are so useful in molecular developmental biology. Of course, such molecular tools are also useful for descriptive purposes (e.g., for measuring the expression of genes in different tissues or for tracking cell lineages).

But other types of causes, too, turned out to be invaluable for learning more about developmental processes. For example, the morphogens (Section 3) led researchers to other components of signaling pathways as well as to other morphogens and target genes. Various selector genes, in particular the Hox genes (Section 4.3), turned out to be highly conserved in animal evolution, presumably because they are *generatively entrenched* (Wimsatt, 2007: Chapter 7) in the developmental systems of all animals. This was an extremely useful property, for it allowed researchers to "fish" for homologous genes using DNA from *Drosophila* (where they were first isolated) as molecular probes. Furthermore, we have also seen that idealized causal models such as the French Flag Model and the associated idealized conception of morphogen were heuristically helpful for biologists in order to formulate specific research questions and for identifying assumptions that were in need of correction.

By highlighting the heuristic value of various kinds of causal knowledge, the causal perspective makes an important contribution to explaining developmental biology's epistemic success before and after the molecular turn.

The third aspect on which a causal perspective shines light is the *centrality of the molecular level*. It is widely assumed that there is something unique about the molecular level in terms of its explanatory power, some kind of fundamentality, perhaps due to the fact that molecules are physical-chemical entities the knowledge of which could connect biological phenomena to the allegedly fundamental theories of physics and chemistry. Much of the debate about reduction and reductionism was motivated by such concerns (see Section 1.2). However, while there are examples where biologists were able to establish such connections, for example, in models of neural excitation (Weber, 2005: Chapter 2) or in the model of self-regulation of morphogen gradients (see Section 3.4), direct physical-chemical explanations in biology still make up only a small fraction of the vast body of knowledge of current biology. Furthermore, many of them are highly idealized or just how-possibly explanations (Section 3.4).

My analyses thus confirm Woodward's (2003) contention that the explanatory importance of the molecular level in recent biology comes not from any metaphysical fundamentality, but from the possibilities of precise experimental interventions that it offers. In developmental biology, much of the molecular knowledge that has been accumulated in recent decades is knowledge about the effects of experimental manipulations on developmental processes. Molecules in particular genes, RNAs and proteins (including antibodies that recognize specific proteins) have proven to be powerful targets for experimentally manipulating developmental processes in model organisms in order to learn about further potential targets for intervention.

The progress of developmental biology turns out to be above all an increase in the manipulability of developmental processes by experimental practices, and molecules are often but not always the keys to successful experimental interventions. Thus, molecular developmental biology has not given us some knowledge that is more fundamental in some metaphysical sense than the knowledge of classical experimental embryology. It's basically the same kind of knowledge – knowledge about the results of experimental interventions – just *a lot more of it*.

To be sure, this interventionist construal of the practice of developmental biology runs the risk of unduly privileging causes that are easily accessible with our current experimental technology, in particular recombinant DNA technology and associated molecular biological methods. Indeed, other kinds of causes such as the mechanical forces that push, pull and squeeze the tissues of a developing organism into its shape have received much less attention, even though this has recently begun to change (Davies, 2013). These causes may be less accessible to interventions[32] and might therefore be left out in my causal-interventionist perspective. However, it should be noted that interventionism only requires that experimental interventions on a cause be possible *in principle*; they need not be *technically* possible for us. However, the easy experimental accessibility of genetically coded molecules with our current technology, as well as their own usefulness as tools for experimental intervention, is very probably the reason why genes & Co. have taken center stage in developmental biology. They are certainly not on a par with other developmental factors in this respect. Not yet.

Thus, we must put to rest the old dream that by reflecting on the development of life forms we could discover some deep metaphysical secrets about nature, something like the existence of vital forces or emergent properties. The remarkable success story of developmental biology over the last 100 years or so may reveal rather more about ourselves than about the metaphysics of nature; about the kinds of causes that appeal to our understanding and our remarkable technical prowess in manipulating things.

---

[32] Thanks to an anonymous reviewer for pointing this out.

References

Anderson, W. 2020. The Compatibility of Differential Equations and Causal Models Reconsidered. *Erkenntnis*, 85, 317–332.

Ankeny, R. A. and S. Leonelli 2020. *Model Organisms* (Cambridge: Cambridge University Press).

Artiga, M. 2020. Signals Are Minimal Causes. *Synthese*, doi: https://doi.org/10.1007/s11229-020-02589-0

Ashe, H. L. and J. Briscoe 2006. The Interpretation of Morphogen Gradients. *Development*, 133, 385–394.

Ayala, F. J. 1974. Introduction, in F. J. Ayala, and T. Dobzhanski (eds.). *Studies in the Philosophy of Biology. Reduction and Related Problems* (Berkeley: University of California Press) VII–XVI.

Baedke, J. 2020. Mechanisms in Evo-Devo, in L. Nuno de la Rosa, and G. Müller (eds.). *Evolutionary Developmental Biology: A Reference Guide* (Cham: Springer International Publishing) 1–14.

Baedke, J. 2021. The Origin of New Levels of Organization, in D. Brooks, J. DiFrisco, and W. C. Wimsatt (eds.). *Levels of Organization in the Biological Sciences* (Cambridge, MA: Massachusetts Institute of Technology Press).

Baetu, T. 2019. *Mechanisms in Molecular Biology* (Cambridge: Cambridge University Press).

Baumgartner, M. and L. Casini 2017. An Abductive Theory of Constitution. *Philosophy of Science*, 84, 214–233.

Baxter, J. 2019. How Biological Technology Should Inform the Causal Selection Debate. *Philosophy, Theory, and Practice in Biology*, 11. doi: https://doi.org/10.3998/ptpbio.16039257.0011.002

Ben-Zvi, D., B.-Z. Shilo, A. Fainsod and N. Barkai 2008. Scaling of the BMP Activation Gradient in Xenopus Embryos. *Nature*, 453, 1205–1211.

Bich, L., M. Mossio, K. Ruiz-Mirazo and A. Moreno 2016. Biological Regulation: Controlling the System from Within. *Biology and Philosophy*, 31, 237–265.

Bourrat, P. 2019. On Calcott's Permissive and Instructive Cause Distinction. *Biology and Philosophy*, 341 (1). doi: https://doi.org/10.1007/s10539-018-9654-y

Braillard, P.-A. and C. Malaterre (eds.) 2015. *Explanation in Biology: An Enquiry into the Diversity of Explanatory Patterns in the Life Sciences* (Dordrecht: Springer).

Brandon, R. N. 1990. *Adaptation and Environment* (Princeton: Princeton University Press).

Brigandt, I. and A. C. Love 2017. Reductionism in Biology, in E. N. Zalta (ed.). *The Stanford Encyclopedia of Philosophy*, accessed at https://plato .stanford.edu/archives/spr2017/entries/reduction-biology/.

Briscoe, J. and D. Small 2015. Morphogen Rules: Design Principles of Gradient-Mediated Embryo Patterning. *Development*, 142, 3996–4009.

Calcott, B. 2017. Causal Specificity and the Instructive–Permissive Distinction. *Biology and Philosophy* 32, 481–505.

Cho, K. W. Y., B. Blumberg, H. Steinbeisser and E. M. De Robertis 1991. Molecular Nature of Spemann's Organizer: The Role of the Xenopus Homeobox Gene *goosecoid*. *Cell*, 67, 1111–1120.

Craver, C. 2007. *Explaining the Brain: Mechanisms and the Mosaic Unity of Neuroscience* (Oxford: Oxford University Press).

Craver, C. and D. M. Kaplan 2020. Are More Details Better? On the Norms of Completeness for Mechanistic Explanations. *British Journal for the Philosophy of Science*, 71, 287–319.

Craver, C., S. Glennan and M. Povich 2021. Constitutive Relevance and Mutual Manipulability Revisited. *Synthese*, doi: https://doi.org/10.1007/s11229-021-03183-8

Davidson, E. H. 1993. Later Embryogenesis: Regulatory Circuitry in Morphogenetic Fields. *Development*, 118, 665–690.

Davies, J. A. 2013. *Mechanisms of Morphogenesis: The Creation of Biological Form* (Amsterdam: Academic Press, Elsevier).

De Chadarevian, S. 1998. Of Worms and Programmes: *Caenorhabditis elegans* and the Study of Development. *Studies in History and Philosophy of Biological and Biomedical Sciences*, 29C 1, 81–106.

De Robertis, E. M. 2006. Spemann's Organizer and Self-Regulation in Amphibian Embryos. *Nature Reviews Molecular Cell Biology*, 7, 296–302.

De Robertis, E. M. and Y. Moriyama 2016. The Chordin Morphogenetic Pathway. *Current Topics in Developmental Biology*, 116, 231–245.

Driesch, H. 1905. *Der Vitalismus als Geschichte und als Lehre* (Leipzig: Barth).

Driesch, H. 1928. *Philosophie des Organischen*, 4th ed. (Leipzig: Quelle & Meyer).

Driever, W. and C. Nüsslein-Volhard 1988. The bicoid Protein Determines Position in the Drosophila Embryo in a Concentration-Dependent Manner. *Cell*, 54, 54–104.

Dubuis, J. O., G. Tkacik, E. F. Wieschaus, T. Gregor and W. Bialek 2013. Positional Information, in Bits. *PNAS*, 110, 16301–16308.

Dupré, J. 2013. I—Living Causes. *Aristotelian Society Supplementary Volume* 87, 19–37.

Dupré, J. 2021. *The Metaphysics of Biology* (Cambridge: Cambridge University Press).

Feest, U. 2010. Concepts as Tools in the Experimental Generation of Knowledge in Cognitive Neuropsychology. *Spontaneous Generations*, 4, 173–190.

Feyerabend, P. K. 1962. Explanation, Reduction and Empiricism, in H. Feigl and G. Maxwell (eds.). *Scientific Explanation, Space, and Time*, Minnesota Studies in the Philosophy of Science, vol. III (Minneapolis: University of Minnesota Press) 28–97.

Franklin-Hall, L. R. 2015. Explaining Causal Selection with Explanatory Causal Economy: Biology and Beyond, in P. A. Braillard and C. Malaterre (eds.). *Explanation in Biology: An Enquiry into the Diversity of Explanatory Patterns in the Life Sciences* (Dordrecht: Springer), 413–438.

García-Bellido, A. 1975. Genetic Control of Wing Disc Development in Drosophila. *Ciba Foundation Symposium*, 29, 161–182.

Gärdenfors, P. and M. Lombard 2020. Technology Led to More Abstract Causal Reasoning. *Biology and Philosophy*, 35 (40). doi: https://doi.org/10.1007/s10539-020-09757-z

Gehring, W. J. 1998. *Master Control Genes in Development and Evolution: The Homeobox Story* (New Haven, CT: Yale University Press).

Gilbert, S. F. 2001. Continuity and Change: Paradigm Shifts in Neural Induction. *The International Journal of Developmental Biology*, 45, 155–164.

Gilbert, S. F. and M. J. F. Barresi 2016. *Developmental Biology*, 11th ed. (Sunderland, MA: Sinauer Associates).

Godfrey-Smith, P. and K. Sterelny 2016. Biological Information, in E. N. Zalta (ed.). *The Stanford Encyclopedia of Philosophy* (Stanford, CA: Metaphysics Research Lab, Stanford University). Accessed at https://plato.stanford.edu/archives/sum2016/entries/information-biological.

Griffiths, P. E. and R. D. Gray 1994. Developmental Systems and Evolutionary Explanation. *The Journal of Philosophy*, 91, 277–304.

Griffiths, P. E., A. Pocheville, B. Calcott et al. 2015. Measuring Causal Specificity. *Philosophy of Science*, 82, 529–555.

Gurwitsch, A. 1922. Über den Begriff des Embryonalen Feldes. *Archiv für Entwicklungsmechanik der Organismen*, 51, 383–415.

Guss, K. A., C. E. Nelson, A. Hudson, M. E. Kraus and S. B. Carroll 2001. Control of a Genetic Regulatory Network by a Selector Gene. *Science*, 292, 1164–1167.

Halder, G., P. Callaerts and W. J. Gehring 1995. Induction of Ectopic Eyes by Targeted Expression of the *eyeless* Gene in Drosophila. *Science*, 267, 1788–1792.

Hamburger, V. 1988. *The Heritage of Experimental Embryology. Hans Spemann and the Organizer.* (Oxford: Oxford University Press).

Hemmati-Brivanlou, A. and D. Melton 1997. Vertebrate Embryonic Cells Will Become Nerve Cells Unless Told Otherwise. *Cell*, 88, 13–17.

Hopwood, N. 2009. Embryology, in J. V. Pickstone and P. J. Bowler (eds.). *The Cambridge History of Science: Volume 6: The Modern Biological and Earth Sciences* (Cambridge: Cambridge University Press), 285–315.

Hoyningen-Huene, P. 1990. Kuhn's Conception of Incommensurability. *Studies in History and Philosophy of Science*, 21, 481–492.

Huxley, J. S. and G. R. De Beer 1934. *The Elements of Experimental Embryology* (Cambridge: Cambridge University Press).

Jaeger, J., S. Surkova, M. Blagov et al. 2004. Dynamic Control of Positional Information in the Early *Drosophila* Embryo. *Nature*, 430, 368–371.

Jaeger, J. and A. Martinez-Arias 2009. Getting the Measure of Positional Information. *PLOS Biology*, 7 (3). doi: https://doi.org/10.1371/journal.pbio.1000081

Kaiser, M. I. 2015. *Reductive Explanation in the Biological Sciences* (Cham: Springer International Publishing).

Kaplan, D. M. and W. Bechtel 2011. Dynamical Models: An Alternative or Complement to Mechanistic Explanations? *Topics in Cognitive Science*, 3, 438–444.

Kay, L. E. 2000. *Who Wrote the Book of Life? A History of the Genetic Code* (Stanford: Stanford University Press).

Kim, J. 2007. *Physicalism, or Something Near Enough* (Princeton: Princeton University Press).

Knuuttila, T. and M. S. Morgan 2019. De-idealization: No Easy Reversals. *Philosophy of Science*, 86, 641–661.

Kuhn, T. S. 1970. *The Structure of Scientific Revolutions*, 2nd ed. (Chicago: The University of Chicago Press).

Laubichler, M. D. and G. P. Wagner 2001. How Molecular Is Molecular Developmental Biology? A Reply to Alex Rosenberg's Reductionism Redux: Computing the Embryo. *Biology and Philosophy*, 16, 53–68.

Lean, O. 2019. Binding Specificity and Causal Selection in Drug Design. *Philosophy of Science*, 87 (1). doi: https://doi.org/10.1086/706093

Levy, A. 2014. Machine-Likeness and Explanation by Decomposition. *Philosophers' Imprint*, 14 (6).

Levy, A. 2011. Information in Biology: A Fictionalist Account 1. *Noûs*, 45, 640–657.

Lewis, J. 2008. From Signals to Patterns: Space, Time, and Mathematics in Developmental Biology. *Science*, 322, 399–403.

Lohmann, I. 2006. Hox Genes: Realising the Importance of Realisators. *Current Biology*, 16, R988–989.

Love, A. C. 2014. The Erotetic Organization of Developmental Biology, in A. Minelli and T. Pradeu (eds.). *Towards a Theory of Development* (Oxford: Oxford University Press), 33–55.

Love, A. C. 2020a. Developmental Biology, in E. N. Zalta (ed.). *The Stanford Encyclopedia of Philosophy* (Stanford: Metaphysics Research Lab, Stanford University). Accessed at https://plato.stanford.edu/archives/spr2020/entries/biology-developmental.

Love, A. C. 2020b. Positional Information and the Measurement of Specificity. *Philosophy of Science*, 87 (5). doi: https://doi.org/10.1086/710617

Lovegrove, B., S. Simões, M. L. Rivas et al. 2006. Coordinated Control of Cell Adhesion, Polarity, and Cytoskeleton Underlies Hox-Induced Organogenesis in Drosophila. *Current Biology*, 16, 2206–2216.

Machamer, P., L. Darden and C. Craver 2000. Thinking about Mechanisms. *Philosophy of Science*, 67, 1–25.

Maienschein, J. 1991. The Origins of Entwicklungsmechanik, in S. F. Gilbert (ed.). *A Conceptual History of Modern Embryology* (London: Plenum Press), 43–61.

Mangold, O. 1933. Über die Induktionsfähigkeit der verschiedenen Bezirke der Neurula von Urodelen. *Naturwissenschaften*, 21, 761–766.

Mann, R. S. and S. B. Carroll 2002. Molecular Mechanisms of Selector Gene Function and Evolution. *Current Opinion in Genetics and Development*, 12, 592–600.

Massagué, J. 1998. TGF-beta Signal Transduction. *Annual Review of Biochemistry*, 67, 753–791.

Mc Manus, F. 2012. Development and Mechanistic Explanation. *Studies in History and Philosophy of Biological and Biomedical Sciences*, 43, 532–541.

Meyer, R. 2020. Dynamical Causes. *Biology and Philosophy*, 35. doi: https://doi.org/10.1007/s10539-020-09755-1

Nicholson, D. J. 2012. The Concept of Mechanism in Biology. *Studies in History and Philosophy of Biological and Biomedical Sciences*, 43, 152–163.

Nicholson, D. J. and J. Dupré (eds.) 2018. *Everything Flows: Towards a Processual Philosophy of Biology* (Oxford: Oxford University Press).

Nickles, T. 1973. Two Concepts of Intertheoretic Reduction. *The Journal of Philosophy*, 70, 181–220.

Oyama, S. 2000. *The Ontogeny of Information: Developmental Systems and Evolution*, 2nd ed., revised and expanded (Durham: Duke University Press).

Pagès, F. and S. Kerridge 2000. Morphogen Gradients. A Question of Time or Concentration? *Trends in Genetics*, 16, 40–44.

Parkkinen, V.-P. 2014. Developmental Explanation, in M. C. Galavotti, D. Dieks, W. J. Gonzalez, S. Hartmann, T. Uebel and M. Weber (eds.). *New Directions in the Philosophy of Science* (Cham: Springer), 157–172.

Plouhinec, J.-L., L. Zakin, Y. Moriyama and E. M. De Robertis 2013. Chordin Forms a Self-Organizing Morphogen Gradient in the Extracellular Space between Ectoderm and Mesoderm in the Xenopus Embryo. *PNAS*, 110, 20372–20379.

Plutynski, A. 2018. *Explaining Cancer: Finding Order in Disorder* (Oxford: Oxford University Press).

Potochnik, A. 2017. *Idealization and the Aims of Science* (Chicago: University of Chicago Press).

Pradeu, T., L. Laplane, M. Morange, A. Nicoglou and M. Vervoort 2011. The Boundaries of Development. *Biological Theory*, 6, 1–3.

Reversade, B. and E. M. De Robertis 2005. Regulation of ADMP and BMP2/4/7 at Opposite Embryonic Poles Generates a Self-Regulating Morphogenetic Field. *Cell*, 123, 1147–1160.

Rogers, K. W. and A. F. Schier 2011. Morphogen Gradients: From Generation to Interpretation. *Annual Review of Cell and Developmental Biology* 27, 377–407.

Rosenberg, A. 1997. Reductionism Redux: Computing the Embryo. *Biology and Philosophy*, 12, 445–470.

Rosenberg, A. 2006. *Darwinian Reductionism* (Chicago: The University of Chicago Press).

Rosenberg, A., 2020. *Reduction and Mechanism* (Cambridge: Cambridge University Press).

Ross, L. N. 2018. Causal Selection and the Pathway Concept. *Philosophy of Science*, 85, 551–572.

Ross, L. N. 2020. Causal Concepts in Biology: How Pathways Differ from Mechanisms and Why It Matters. *British Journal for the Philosophy of Science*, 72, 131–158.

Ross, L. N. forthcoming. Causal Control: A Rationale for Causal Selection, in B. Hanley, C. K. Waters and J. Woodward (eds.). *Philosophical Perspectives on Causal Reasoning in Biology*, Minnesota Studies in Philosophy of Science (Minneapolis: University of Minnesota Press).

Roth, S. and J. Lynch 2012. Does the Bicoid Gradient Matter? *Cell*, 149, 511–512.

Saha, M. S. 1991. Spemann Seen through a Lens, in S. F. Gilbert (ed.). *A Conceptual History of Modern Embryology* (London: Plenum Press) 91–108.

Schaffner, K. F. 1993. *Discovery and Explanation in Biology and Medicine*. Chicago: University of Chicago Press.

Silberstein, M. and A. Chemero 2013. Constraints on Localization and Decomposition as Explanatory Strategies in the Biological Sciences. *Philosophy of Science*, 80, 958–970.

Skipper, R. A. and R. Millstein 2005. Thinking about Evolutionary Mechanisms: Natural Selection. *Studies in History and Philosophy of Biological and Biomedical Sciences*, 36, 327–347.

Skyrms, B. 2010. *Signals. Evolution, Learning, & Information.* (Oxford: Oxford University Press).

Slack, J. M. W. 1993. Embryonic Induction. *Mechanisms of Development*, 41, 91–107.

Sober, E. 1999. The Multiple Realizability Argument against Reductionism. *Philosophy of Science*, 66, 542–564.

Spemann, H. 1936. *Experimentelle Beiträge zu einer Theorie der Entwicklung* (Berlin: Julius Springer).

Stepp, N., A. Chemero and M. T. Turvey 2011. Philosophy for the Rest of Cognitive Science. *Topics in Cognitive Science*, 3, 425–437.

Waddington, C. H. 1956. *Principles of Embryology* (London: Allen and Unwin).

Waters, C. K. 2007. Causes That Make a Difference. *The Journal of Philosophy*, 104, 551–579.

Waters, C. K. 2008. Beyond Theoretical Reduction and Layer-Cake Antireduction: How DNA Retooled Genetics and Transformed Biological Practice, in M. Ruse (ed.). *The Oxford Handbook of Philosophy of Biology* (Oxford: Oxford University Press).

Waters, C. K. 2017. No General Structure, in M. H. Slater and Z. Yudell (eds.). *Metaphysics and the Philosophy of Science: New Essays* (Oxford: Oxford University Press) 81–108.

Weber, M. 1999. Hans Drieschs Argumente für den Vitalismus. *Philosophia Naturalis*, 36, 265–295.

Weber, M. 2005. *Philosophy of Experimental Biology* (Cambridge: Cambridge University Press).

Weber, M. 2006. The Central Dogma as a Thesis of Causal Specificity. *History and Philosophy of the Life Sciences*, 28, 565–580.

Weber, M., 2008. Causes without Mechanisms: Experimental Regularities, Physical Laws, and Neuroscientific Explanation. *Philosophy of Science*, 75, 995–1007.

Weber, M. 2016. On the Incompatibility of Dynamical Biological Mechanisms and Causal Graphs. *Philosophy of Science*, 83, 959–971.

Weber, M. forthcoming-a. Causal Selection versus Causal Parity in Biology: Relevant Counterfactuals and Biologically Normal Interventions, in B. Hanley, C. K. Waters and J. Woodward (eds.), *Philosophical Perspectives on Causal Reasoning in Biology*, Minnesota Studies in Philosophy of Science (Minneapolis: University of Minnesota Press).

Weber, M., forthcoming-b. A Tale of Three Sciences. From Reduction to Inter-level Practices in Developmental Biology, in W. Bausman, J. Baxter, O. Lean, A. Love and C. K. Waters (eds.). *From Biological Practice to Scientific Metaphysics*, Minnesota Studies in Philosophy of Science (Minneapolis: University of Minnesota Press).

Weisberg, M. 2013. *Simulation and Similarity: Using Models to Understand the World.* (Oxford: Oxford University Press).

Wessells, N. K. 1977. *Tissue Interactions and Development* (Menlo Park, CA: W. A. Benjamin).

Wilkenfeld, D. A. 2013. Understanding as Representation Manipulability. *Synthese*, 190, 997–1016.

Wimsatt, W. C. 2007. *Re-engineering Philosophy for Limited Beings: Piecewise Approximations to Reality* (Cambridge, MA: Harvard University Press).

Wolpert, L. 1969. Positional Information and the Spatial Pattern of Cellular Differentiation. *Journal of Theoretical Biology*, 25, 1–47.

Woodward, J. 2003. *Making Things Happen: A Theory of Causal Explanation* (Oxford: Oxford University Press).

Woodward, J. 2010. Causation in Biology: Stability, Specificity, and the Choice of Levels of Explanation. *Biology and Philosophy*, 25, 287–318.

Woodward, J. 2013. II – Mechanistic Explanation: Its Scope and Limits. *Aristotelian Society Supplementary Volume* 87, 39–65.

Woodward, J. and L. Ross 2021. Scientific Explanation, in E. N. Zalta (ed.). *The Stanford Encyclopedia of Philosophy* (Stanford: Metaphysics Research Lab, Stanford University). Accessed at https://plato.stanford.edu/archives/sum2021/entries/scientific-explanation/

Woody, A. I. 2015. Re-orienting Discussions of Scientific Explanation: A Functional Perspective. *Studies in History and Philosophy of Science*, 52, 79–87.

Acknowledgments

This Element was conceived during a sabbatical at the Department of Logic and Philosophy of Science at the University of California, Irvine in Fall 2019. I am indebted to this Department for its generous support, the kind hospitality and for providing a highly stimulating research environment. I wish to acknowledge in particular helpful suggestions from Lauren Ross, Simon Huttegger, Brian Skyrms and Jeff Barrett, as well as from Jim Woodward and Bill Bechtel. My research has also benefited from grants from the Swiss National Science Foundation (100017_169810) and the John Templeton Foundation (2015–18). Open access publication is supported by the Swiss National Science Foundation (10BP12_206575). I wish to thank my co-PIs Alan Love, Ken Waters and Bill Wimsatt from the latter project as well as the numerous participants of the three Summer Institutes in Basel, Calgary and Taipei for many stimulating discussions. Some ideas were also presented at the European Advanced School in Philosophy of the Life Sciences 2018 at the KLI in Klosterneuburg, where Elena Rondeau and Naïd Mubalegh gave thoughtful commentaries. Special thanks also to two anonymous reviewers and the members of the lgBIG group in Geneva for helpful comments, in particular María José Ferreira Ruiz, Lorenzo Casini, Silvia de Cesare, Michaela Egli, Michal Hladky, Christian Sachse, Mike Stuart and Will Bausman. I am indebted also to Grant Ramsey and Michael Ruse for their advice and their enthusiasm for this project. Finally, I wish to thank Vera, Lionel and Valentin for their love and understanding.

Cambridge Elements ⁼

Philosophy of Biology

Grant Ramsey
KU Leuven

Grant Ramsey is a BOFZAP research professor at the Institute of Philosophy, KU Leuven, Belgium. His work centers on philosophical problems at the foundation of evolutionary biology. He has been awarded the Popper Prize twice for his work in this area. He also publishes in the philosophy of animal behavior, human nature and the moral emotions. He runs the Ramsey Lab (theramseylab.org), a highly collaborative research group focused on issues in the philosophy of the life sciences.

Michael Ruse
Florida State University

Michael Ruse is the Lucyle T. Werkmeister Professor of Philosophy and the Director of the Program in the History and Philosophy of Science at Florida State University. He is Professor Emeritus at the University of Guelph, in Ontario, Canada. He is a former Guggenheim fellow and Gifford lecturer. He is the author or editor of over sixty books, most recently *Darwinism as Religion: What Literature Tells Us about Evolution; On Purpose; The Problem of War: Darwinism, Christianity, and their Battle to Understand Human Conflict;* and *A Meaning to Life.*

About the Series

This Cambridge Elements series provides concise and structured introductions to all of the central topics in the philosophy of biology. Contributors to the series are cutting-edge researchers who offer balanced, comprehensive coverage of multiple perspectives, while also developing new ideas and arguments from a unique viewpoint.

Cambridge Elements ⹅

Philosophy of Biology

Elements in the Series

A full series listing is available at www.cambridge.org/EPBY